Student Lab Manual

for

Argument-Driven Inquiry

in

LIFE SCIENCE

LAB INVESTIGATIONS
for GRADES 6–8

Student Lab Manual

for
Argument-Driven Inquiry
in
LIFE SCIENCE

LAB INVESTIGATIONS
for GRADES 6–8

Patrick J. Enderle, Ruth Bickel, Leeanne Gleim, Ellen Granger,
Jonathon Grooms, Melanie Hester, Ashley Murphy, Victor Sampson,
and Sherry A. Southerland

NSTApress
National Science Teachers Association
Arlington, Virginia

National Science Teachers Association

Claire Reinburg, Director
Wendy Rubin, Managing Editor
Rachel Ledbetter, Associate Editor
Amanda O'Brien, Associate Editor
Donna Yudkin, Book Acquisitions Coordinator

ART AND DESIGN
Will Thomas Jr., Director

PRINTING AND PRODUCTION
Catherine Lorrain, Director

NATIONAL SCIENCE TEACHERS ASSOCIATION
David L. Evans, Executive Director
David Beacom, Publisher

1840 Wilson Blvd., Arlington, VA 22201
www.nsta.org/store
For customer service inquiries, please call 800-277-5300.

FSC
www.fsc.org
MIX
Paper from
responsible sources
FSC® C011935

Cataloging-in-Publication Data are available from the Library of Congress.
    LCCN: 2015013671

CONTENTS

Acknowledgments ix
About the Authors xi

SECTION 1
Introduction and Lab Safety

Introduction by Patrick J. Enderle and Victor Sampson 3

Safety in the Science Classroom, Laboratory, or Field Sites 5

SECTION 2—Life Sciences Core Idea 1
From Molecules to Organisms: Structures and Processes

INTRODUCTION LABS

Lab 1. Cellular Respiration: Do Plants Use Cellular Respiration to Produce Energy?
Lab Handout 14
Checkout Questions 19

Lab 2. Photosynthesis: Where Does Photosynthesis Take Place in Plants?
Lab Handout 21
Checkout Questions 26

APPLICATION LABS

Lab 3. Osmosis: How Does the Concentration of Salt in Water Affect the Rate of Osmosis?
Lab Handout 30
Checkout Questions 35

Lab 4. Cell Structure: What Type of Cell Is on the Unknown Slides?
Lab Handout 37
Checkout Questions 42

Lab 5. Temperature and Photosynthesis: How Does Temperature Affect the Rate of Photosynthesis in Plants?
Lab Handout 44
Checkout Questions 49

Lab 6. Energy in Food: Which Type of Nut Is Best for a New Energy Bar?
Lab Handout 51
Lab 6 Reference Sheet: Costs and Exercise Calories 56
Checkout Questions 57

Lab 7. Respiratory and Cardiovascular Systems: How Do Activity and Physical Factors Relate to Respiratory and Cardiovascular Fitness?

Lab Handout.. 59

Lab 7 Reference Sheet: Cardiovascular Fitness Test Protocol and Tables.......................... 64

Checkout Questions.. 68

Lab 8. Memory and Stimuli: How Does the Way Information Is Presented Affect Working Memory?

Lab Handout.. 70

Checkout Questions.. 75

SECTION 3—Life Sciences Core Idea 2
Ecosystems: Interactions, Energy, and Dynamics

INTRODUCTION LABS

Lab 9. Population Growth: What Factors Limit the Size of a Population of Yeast?

Lab Handout.. 80

Checkout Questions.. 86

Lab 10. Predator-Prey Relationships: How Is the Size of a Predator Population Related to the Size of a Prey Population?

Lab Handout.. 88

Checkout Questions.. 93

APPLICATION LABS

Lab 11. Food Webs and Ecosystems: Which Member of an Ecosystem Would Affect the Food Web the Most If Removed?

Lab Handout.. 96

Checkout Questions.. 100

Lab 12. Matter in Ecosystems: How Healthy Are Your Local Ecosystems?

Lab Handout.. 102

Lab 12 Reference Sheet: The Nitrogen Cycle and the Phosphorus Cycle.......................... 107

Checkout Questions.. 110

Lab 13. Carbon Cycling: Which Carbon Cycle Process Affects Atmospheric Carbon the Most?

Lab Handout.. 112

Checkout Questions.. 117

SECTION 4—Life Sciences Core Idea 3
Heredity: Inheritance and Variation in Traits

INTRODUCTION LABS

Lab 14. Variation in Traits: How Do Beetle Traits Vary Within and Across Species?

Lab Handout.. 122
Lab 14 Reference Sheet: Three Types of Beetles 127
Checkout Questions.. 130

Lab 15. Mutations in Genes: How Do Different Types of Mutations in Genes Affect the Function of an Organism?

Lab Handout.. 132
Checkout Questions.. 137

APPLICATION LAB

Lab 16. Mechanisms of Inheritance: How Do Fruit Flies Inherit the Sepia Eye Color Trait?

Lab Handout.. 140
Checkout Questions.. 145

SECTION 5—Life Sciences Core Idea 4
Biological Evolution: Unity and Diversity

INTRODUCTION LAB

Lab 17. Mechanisms of Evolution: Why Does a Specific Version of a Trait Become More Common in a Population Over Time?

Lab Handout.. 150
Checkout Questions.. 156

APPLICATION LABS

Lab 18. Environmental Change and Evolution: Which Mechanism of Microevolution Caused the Beak of the Medium Ground Finch Population on Daphne Major to Increase in Size From 1976 to 1978?

Lab Handout.. 160
Checkout Questions.. 166

Lab 19. Phylogenetic Trees and the Classification of Fossils: How Should Biologists Classify the Seymouria?
> Lab Handout.. 169
>
> Checkout Questions.. 176

Lab 20. Descent With Modification and Embryonic Development: Does Animal Embryonic Development Support or Refute the Theory of Descent With Modification?
> Lab Handout.. 178
>
> Checkout Questions.. 184
>
> Image Credits 187

ACKNOWLEDGMENTS

The development of this book was supported by the Institute of Education Sciences, U.S. Department of Education, through grant R305A100909 to Florida State University. The opinions expressed are those of the authors and do not represent the views of the institute or the U.S. Department of Education.

ABOUT THE AUTHORS

Patrick J. Enderle is a research faculty member in the Center for Education Research in Mathematics, Engineering, and Science (CERMES) at The University of Texas at Austin (UT-Austin). He received his BS and MS in molecular biology from East Carolina University. Patrick then spent some time as a high school biology teacher and several years as a visiting professor in the Department of Biology at East Carolina University. He then attended Florida State University (FSU), where he graduated with a PhD in science education. His research interests include argumentation in the science classroom, science teacher professional development, and enhancing undergraduate science education. To learn more about his work in science education, go to *http://patrickenderle.weebly.com*.

Ruth Bickel has been a teacher at FSU Schools for several years, supporting student learning in a variety of disciplines. She was originally a social studies teacher before taking an interest in teaching science. She has taught middle school Earth and space science and life science for several years. She has also taught a high school–level forensics course over the past few years. Ruth was responsible for writing and piloting many of the lab investigations included in this book.

Leeanne Gleim received a BA in elementary education from the University of Southern Indiana and an MS in science education from FSU. While at FSU, she worked as a research assistant for Victor Sampson (see his biography later in this section). After graduating, she taught biology and honors biology at FSU Schools, where she participated in the development of the argument-driven inquiry model. Leeanne was also responsible for writing and piloting many of the lab investigations included in this book.

Ellen Granger is the director of the Office of Science Teaching Activities and co-director of FSU-Teach, a collaborative math and science teacher preparation program between the College of Arts and Sciences and the College of Education at FSU. She earned her doctorate in neuroscience from FSU. She is a practicing scientist and science educator and has worked in teacher professional development for almost 20 years. In November 2013, she was named a Fellow of the American Association for the Advancement of Science for "distinguished contribution, service and leadership in advancing knowledge and classroom practices in science education."

Jonathon Grooms received a BS in secondary science and mathematics teaching with a focus in chemistry and physics from FSU. Upon graduation, Jonathon joined FSU's Office of Science Teaching Activities, where he directed the physical science outreach program Science on the Move. He entered graduate school at FSU and earned a PhD in science education. He now serves as a research scientist in CERMES (Center for Education Research in Mathematics, Engineering, and Science) at FSU. To learn more about his work in science education, go to *www.jgrooms.com*.

Melanie Hester has a BS in biological sciences with minors in chemistry and classical civilizations from FSU and an MS in secondary science education from FSU. She has been teaching for more than 20 years, with the last 13 at the FSU School in Tallahassee. Melanie was a Lockheed Martin fellow and a Woodrow Wilson fellow and received a Teacher of the Year award in 2007. She frequently gives presentations about innovative approaches to teaching at conferences and works with preservice teachers. Melanie was responsible for writing and piloting many of the lab investigations included in this book.

Ashley Murphy attended FSU and earned a BS with dual majors in biology and secondary science education. Ashley spent some time as a middle school biology and science teacher before entering graduate school at UT-Austin, where she is currently working toward a PhD in STEM (science, technology, engineering, and mathematics) education. Her research interests include argumentation in elementary and middle school classrooms. As an educator, she frequently employed argumentation as a means to enhance student understanding of concepts and science literacy.

Victor Sampson is an associate professor of science education and the director of CERMES at UT-Austin. He received a BA in zoology from the University of Washington, an MIT from Seattle University, and a PhD in curriculum and instruction with a specialization in science education from Arizona State University. Victor taught high school biology and chemistry for nine years before taking a position at FSU and then moving to UT-Austin. He specializes in argumentation in science education, teacher learning, and assessment. To learn more about his work in science education, go to *www.vicsampson.com*.

Sherry A. Southerland is a professor at FSU and the co-director of FSU-Teach. She received a BS and an MS in biology from Auburn University and a PhD in curriculum and instruction from Louisiana State University, with a specialization in science education and evolutionary biology. Sherry has worked as a teacher educator, biology instructor, high school science teacher, field biologist, and forensic chemist. Her research interests include understanding the influence of culture and emotions on learning—specifically evolution education and teacher education—and understanding how to better support teachers in shaping the way they approach science teaching and learning.

SECTION 1
Introduction and Lab Safety

INTRODUCTION

By Patrick J. Enderle and Victor Sampson

Science is more than a collection of facts or ideas that describe what we know about how the world works and why it works that way. Science is also a set of crosscutting concepts and practices that people can use to develop and refine new explanations for, or descriptions of, the natural world. These core ideas, crosscutting concepts, and practices of science are important for you to learn. When you understand these, it is easier to appreciate the beauty and wonder of science, to engage in public discussions about science, and to evaluate the strengths of scientific information presented through popular media. You will also have the knowledge and skills needed to continue learning science outside school or to enter a career in science, engineering, or technology.

The core ideas of science—based on evidence from many investigations—include theories, laws, and models that scientists use to explain natural events and bodies of data and to predict the results of new investigations. The crosscutting concepts are themes that have value in all areas of science and are used to help us understand a natural phenomenon. They can be used to connect knowledge from the various fields of science into a coherent and scientifically based view of the world. Finally, the practices of science are used to develop and refine new ideas about the world. Although some practices are specific to certain fields of science, all fields share a set of common practices. The practices include such things as asking and answering questions; planning and carrying out investigations; analyzing and interpreting data; and obtaining, evaluating, and communicating information. One of the most important practices of science is arguing from evidence. Arguing from evidence involves proposing, supporting, challenging, and refining claims based on evidence. Arguing is important because scientists need to be able to examine, review, and evaluate their own ideas and to critique those of others. Scientists also argue from evidence when they need to judge the quality of data, produce and improve models, develop new questions from those models that can be investigated, and suggest ways to refine or modify existing theories, laws, and models.

Always remember that science is a social activity, not an individual one. Science is social because many different scientists contribute to the development of new scientific knowledge. As scientists carry out their research, they frequently talk with their colleagues, both formally, like at a meeting, and informally, like in a hallway. They exchange emails, engage in discussions at conferences, share research techniques and analytical procedures, and present new ideas by writing articles in journals or chapters in books. They also critique the ideas and methods used by other scientists through a formal peer review process before they can be published in journals or books. In short, scientists are members of a community, the members of which work together to build, develop, test, critique, and refine ideas. The ways scientists talk, write, think, and interact with each other reflect common ideas about what counts as quality and shared standards for how new ideas should be developed, shared, evaluated, and refined. These ways of interacting make science different from other ways of knowing. The core ideas, crosscutting concepts, and practices of science are important among scientists because most, if not all, scientists find them to be a useful in developing and refining new explanations for, or descriptions of, the natural world.

The laboratory investigations included in this book are designed to help you learn the core ideas, crosscutting concepts, and practices of science. During each investigation, you will have an opportunity to use a core idea, several crosscutting concepts, and the practices of science to understand a natural phenomenon or solve a problem. Your teacher will introduce each investigation by giving you a task to accomplish and a guiding question to answer. You will then work as part of a team to plan and carry out an investigation to collect the data you need to answer that question. From there, your team will develop an initial argument that includes a claim, evidence in support of your claim, and a justification of your evidence. The claim will be your answer to the guiding question, the evidence will include your analysis of the data you collected and an interpretation of that analysis, and the justification will explain why your evidence is important in terms of key science concepts. Next, you will have an opportunity to share your argument with your classmates and to critique their arguments, much like professional scientists do. You will then revise your initial argument based on your colleagues' feedback. Finally, you will be asked to write an investigation report on your own to share what you learned. The report will go through double-blind peer review so you can improve it before you submit it to you teacher for a grade. As you complete more and more investigations in this lab manual, you will not only learn the core ideas associated with each investigation but also get better at using the crosscutting concepts and practices of science to understand the natural world.

SAFETY IN THE SCIENCE CLASSROOM, LABORATORY, OR FIELD SITES

Note to science teachers and supervisors/ administrators: The following safety acknowledgment form is for your use in the classroom and should be given to students at the beginning of the school year to help them understand their role in ensuring a safer and productive science experience.

Science is a process of discovering and exploring the natural world. Exploration occurs in the classroom/laboratory or in the field. As part of your science class, you will be doing many activities and investigations that will involve the use of various materials, equipment, and chemicals. Safety in the science classroom, laboratory, or field sites is the FIRST PRIORITY for students, instructors, and parents. To ensure safer classroom/laboratory/field experiences, the following **Science Rules and Regulations** have been developed for the protection and safety of all. Your instructor will provide additional rules for specific situations or settings. The rules and regulations must be followed at all times. After you have reviewed them with your instructor, read and review the rules and regulations with your parent/guardian. Their signature and your signature on the safety acknowledgment form are required before you will be permitted to participate in any activities or investigations. Your signature indicates that you have read these rules and regulations, understand them, and agree to follow them at all times while working in the classroom/laboratory or in the field.

Source: National Science Teachers Association (NSTA). Safety in the Science Classroom. *www.nsta.org/pdfs/SafetyInTheScienceClassroom.pdf.*

Safety Standards of Student Conduct in the Classroom, Laboratory, and in the Field

1. Conduct yourself in a responsible manner at all times. Frivolous activities, mischievous behavior, throwing items, and conducting pranks are prohibited.

2. Lab and safety information and procedures must be read ahead of time. All verbal and written instructions shall be followed in carrying out the activity or investigation.

3. Eating, drinking, gum chewing, applying cosmetics, manipulating contact lenses, and other unsafe activities are not permitted in the laboratory.

4. Working in the laboratory without the instructor present is prohibited.

5. Unauthorized activities or investigations are prohibited. Unsupervised work is not permitted.

6. Entering preparation or chemical storage areas is prohibited at all times.

7. Removing chemicals or equipment from the classroom or laboratory is prohibited unless authorized by the instructor.

Personal Safety

8. Sanitized indirectly vented chemical splash goggles or safety glasses as appropriate (meeting the ANSI Z87.1 standard) shall be worn during activities or demonstrations in the classroom, laboratory, or field, including pre-laboratory work and clean-up, unless the instructor specifically states that the activity or demonstration does not require the use of eye protection.

9. When an activity requires the use of laboratory aprons, the apron shall be appropriate to the size of the student and the hazard associated with the activity or investigation. The apron shall remain tied throughout the activity or investigation.

10. All accidents, chemical spills, and injuries must be reported immediately to the instructor, no matter how trivial they may seem at the time. Follow your instructor's directions for immediate treatment.

11. Dress appropriately for laboratory work by protecting your body with clothing and shoes. This means that you should use hair ties to tie back long hair and tuck into the collar. Do not wear loose or baggy clothing or dangling jewelry on laboratory days. Acrylic nails are also a safety hazard near heat sources and should not be used. Sandals or open-toed shoes are not to be worn during any lab activities. Refer to pre-lab instructions. If in doubt, ask!

12. Know the location of all safety equipment in the room. This includes eye wash stations, the deluge shower, fire extinguishers, the fume hood, and the safety blanket. Know the location of emergency master electric and gas shut offs and exits.

13. Certain classrooms may have living organisms including plants in aquaria or other containers. Students must not handle organisms without specific instructor authorization. Wash your hands with soap and water after handling organisms and plants.

14. When an activity or investigation requires the use of laboratory gloves for hand protection, the gloves shall be appropriate for the hazard and worn throughout the activity.

Specific Safety Precautions Involving Chemicals and Lab Equipment

15. Avoid inhaling in fumes that may be generated during an activity or investigation.

16. Never fill pipettes by mouth suction. Always use the suction bulbs or pumps.

17. Do not force glass tubing into rubber stoppers. Use glycerin as a lubricant and hold the tubing with a towel as you ease the glass into the stopper.

18. Proper procedures shall be followed when using any heating or flame producing device especially gas burners. Never leave a flame unattended.

19. Remember that hot glass looks the same as cold glass. After heating, glass remains hot for a very long time. Determine if an object is hot by placing your hand close to the object but do not touch it.

20. Should a fire drill, lockdown, or other emergency occur during an investigation or activity, make sure you turn off all gas burners and electrical equipment. During an evacuation emergency, exit the room as directed. During a lockdown, move out of the line of sight from doors and windows if possible or as directed.

21. Always read the reagent bottle labels twice before you use the reagent. Be certain the chemical you use is the correct one.

22. Replace the top on any reagent bottle as soon as you have finished using it and return the reagent to the designated location.

23. Do not return unused chemicals to the reagent container. Follow the instructor's directions for the storage or disposal of these materials.

Standards For Maintaining a Safer Laboratory Environment

24. Backpacks and books are to remain in an area designated by the instructor and shall not be brought into the laboratory area.

25. Never sit on laboratory tables.

26. Work areas should be kept clean and neat at all times. Work surfaces are to be cleaned at the end of each laboratory or activity.

27. Solid chemicals, metals, matches, filter papers, broken glass, and other materials designated by the instructor are to be deposited in the proper waste containers, not in the sink. Follow your instructor's directions for disposal of waste.

28. Sinks are to be used for the disposal of water and those solutions designated by the instructor. Other solutions must be placed in the designated waste disposal containers.

29. Glassware is to be washed with hot, soapy water and scrubbed with the appropriate type and sized brush, rinsed, dried, and returned to its original location.

30. Goggles are to be worn during the activity or investigation, clean up, and through hand washing.

31. Safety Data Sheets (SDSs) contain critical information about hazardous chemicals of which students need to be aware. Your instructor will review the salient points on the SDSs for the hazardous chemicals students will be working with and also post the SDSs in the lab for future reference.

Safety Acknowledgment Form: Science Rules and Regulations

I have read the science rules and regulations in the *Student Lab Manual for Argument-Driven Inquiry in Life Science,* and I agree to follow them during any science course, investigation, or activity. By signing this form, I acknowledge that the science classroom, laboratory, or field sites can be an unsafe place to work and learn. The safety rules and regulations are developed to help prevent accidents and to ensure my own safety and the safety of my fellow students. I will follow any additional instructions given by my instructor. I understand that I may ask my instructor at any time about the rules and regulations if they are not clear to me. My failure to follow these science laboratory rules and regulations may result in disciplinary action.

_____   _____
Student Signature   Date

_____   _____
Parent/Guardian Signature   Date

National Science Teachers Association

SECTION 2
Life Sciences
Core Idea 1

From Molecules to Organisms: Structures and Processes

Introduction Labs

Lab Handout

Lab 1. Cellular Respiration: Do Plants Use Cellular Respiration to Produce Energy?

Introduction

One characteristic of living things is they must take in nutrients and give off waste in order to survive. This is because all living tissues (which are made of cells) are constantly using energy. In animals, this energy comes from a reaction called *cellular respiration*. Cellular respiration refers to a process that occurs inside cells where sugar is used as a fuel source. This process happens in a specific location inside cells called the *mitochondrion* (the plural form is *mitochondria*). Figure L1.1 shows a drawing and an image from an electron microscope of a mitochondrion. Mitochondria are found in both plant and animal cells.

FIGURE L1.1

(a) Drawing and (b) electron microscopic image of a mitochondrion

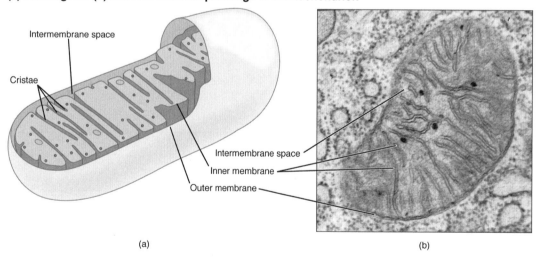

(a)

(b)

The energy that cells use comes from the chemical bonds in sugar. During cellular respiration, oxygen helps convert the chemical energy in sugar molecules into a form animals can use. The energy is transferred by moving electrons from one molecule to another. When electrons are added or taken away, new chemical bonds and types of molecules can be formed. The oxygen helps transfer electrons from sugar to the chemical bonds in another molecule. Most living organisms use a special molecule known as ATP to provide energy for all the activities taking place in their cells. The following equation describes this process:

Sugar ($C_6H_{12}O_6$) + oxygen (O_2) → water (H_2O) + carbon dioxide (CO_2) + usable energy (ATP)

We know that humans use this process to produce energy because when a human breathes, the air that he or she inhales contains about 21% O_2 and less than 1% CO_2; however, when he or she exhales, the air contains about 15% O_2 and 5% CO_2. We also know that all animals use this process to produce energy. It is a unifying characteristic of animals, but what about other types of living things like plants? Do these organisms use this process as well? In this lab investigation you will use an O_2 or CO_2 gas sensor to determine if plants use cellular respiration to produce energy just like animals.

Your Task

Design a scientific investigation to determine if plants use the process of cellular respiration to produce energy. To do this, you will need to use sensors to determine if these organisms cause a change in the CO_2 or O_2 concentrations of air.

The guiding question of this investigation is, **Do plants use cellular respiration to produce energy?**

Materials

You may use any of the following materials during your investigation:

Consumables	Equipment
• Germinating peas (i.e., peas that have been soaked in water) • Dry peas • Plastic beads	• CO_2 or O_2 gas sensor • Biochambers or sealed containers with opening for sensors • Go!Link adaptor and laptop computer • Sanitized indirectly vented chemical-splash goggles • Chemical-resistant apron • Gloves

Safety Precautions

Follow all normal lab safety rules. In addition, take the following safety precautions:

1. Put on sanitized indirectly vented chemical-splash goggles and laboratory apron and gloves before starting the lab activity.

2. Wash hands with soap and water after completing the lab activity.

LAB 1

Getting Started

During your investigation you will need to determine if the peas are producing CO_2 or using O_2. To do this, you can use CO_2 and O_2 gas sensors (see Figure L1.2). To answer the guiding question, you will need to design and conduct an experiment. To accomplish this task, you must first determine what type of data you need to collect, how you will collect it, and how you will analyze the data.

FIGURE L1.2 _____

An O_2 or a CO_2 gas sensor can be used to measure changes in gas concentration.

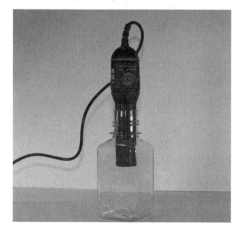

To determine *what type of data you need to collect*, think about the following questions:

- What information will tell you that cellular respiration is occurring in the peas?
- How will the sensors help you measure cellular respiration?
- What type of measurements or observations will you need to record during your investigation?

To determine *how you will collect your data*, think about the following questions:

- What will serve as a control (or comparison) condition?
- What types of treatment conditions will you need to set up and how will you do it?
- How often will you collect data and when will you do it?
- How will you make sure that your data are of high quality (i.e., how will you reduce error)?
- How will you keep track of the data you collect and how will you organize it?

To determine *how you will analyze the data*, think about the following questions:

- How will you determine if there is a difference between the treatment conditions and the control condition?
- What type of calculations will you need to make?
- What type of graph could you create to help make sense of your data?

Connections to Crosscutting Concepts, the Nature of Science, and the Nature of Scientific Inquiry

As you work through your investigation, be sure to think about

- how scientists try to figure out cause-and-effect relationships that explain why something happens,

- how energy and matter flow through living things while being totally conserved,

- how observations and inferences are different but related to each other, and

- the difference between data collected and evidence created in an investigation.

Initial Argument

Once your group has finished collecting and analyzing your data, you will need to develop an initial argument. Your argument must include a claim, evidence to support your claim, and a justification of the evidence. The claim is your group's answer to the guiding question. The evidence is an analysis and interpretation of your data. Finally, the justification of the evidence is why your group thinks the evidence matters. The justification of the evidence is important because scientists can use different kinds of evidence to support their claims. Your group will create your initial argument on a whiteboard. Your whiteboard should include all the information shown in Figure L1.3.

FIGURE L1.3 _____

Argument presentation on a whiteboard

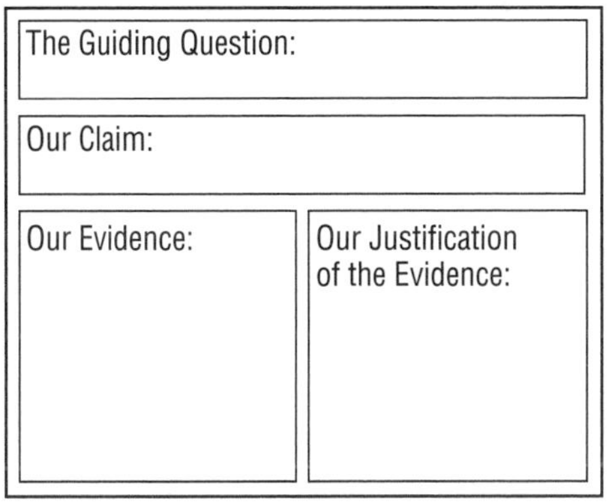

The Guiding Question:	
The Guiding Question:	
Our Claim:	
Our Evidence:	Our Justification of the Evidence:

Argumentation Session

The argumentation session allows all of the groups to share their arguments. One member of each group will stay at the lab station to share that group's argument, while the other members of the group go to the other lab stations one at a time to listen to and critique the arguments developed by their classmates. This is similar to how scientists present their arguments to other scientists at conferences. If you are responsible for critiquing your classmates' arguments, your goal is to look for mistakes so these mistakes can be fixed and they can make their argument better. The argumentation session is also a good time to think about ways you can make your initial argument better. Scientists must share and critique arguments like this to develop new ideas.

To critique an argument, you might need more information than what is included on the whiteboard. You will therefore need to ask the presenter lots of questions. Here are some good questions to ask:

- What did your group do to collect the data? Why do you think that way is the best way to do it?

- What did your group do to analyze the data? Why did your group decide to analyze it that way?

- What other ways of analyzing and interpreting the data did your group talk about?

- What did your group do to make sure that these calculations are correct?
- Why did your group decide to present your evidence in that way?
- What other claims did your group discuss before you decided on that one? Why did your group abandon those other ideas?
- How sure are you that your group's claim is accurate? What could you do to be more certain?

Once the argumentation session is complete, you will have a chance to meet with your group and revise your original argument. Your group might need to gather more data or design a way to test one or more alternative claims as part of this process. Remember, your goal at this stage of the investigation is to develop the most valid or acceptable answer to the research question!

Report

Once you have completed your research, you will need to prepare an investigation report that consists of three sections that provide answers to the following questions:

1. What question were you trying to answer and why?
2. What did you do during your investigation and why did you conduct your investigation in this way?
3. What is your argument?

Your report should answer these questions in two pages or less. The report must be typed, and any diagrams, figures, or tables should be embedded into the document. Be sure to write in a persuasive style; you are trying to convince others that your claim is acceptable or valid!

Checkout Questions

Lab 1. Cellular Respiration: Do Plants Use Cellular Respiration to Produce Energy?

1. Susan and Jessica are having a discussion in science class about the process of cellular respiration. Susan makes the argument that animals use cellular respiration to produce energy based on the evidence that they breathe out carbon dioxide (CO_2); she also said that plants do not use cellular respiration because plants release oxygen (O_2) to the atmosphere. Jessica disagrees with Susan and claims that both plants and animals use cellular respiration. Who do you agree with, Susan or Jessica? Explain your reasoning.

2. Describe the similar structures and features of plant and animal cells that allow for cellular respiration to happen in both types of organisms.

3. In this investigation you measured the amount of gas in the biochamber with a sensor. Were those measurements data or evidence?

    a. Data

    b. Evidence

    c. Unsure

    Explain your answer.

4. Scientists always agree on their observations, but may disagree on the inferences.

   a. I agree with this statement.
   b. I disagree with this statement.

   Explain your answer, using an example from your investigation about cellular respiration.

5. An important goal in science is to develop causal explanations for observations. Explain what a causal explanation is and why it is important, using an example from your investigation about cellular respiration.

6. It is important for scientists to understand the flow of energy in a system. Explain why this is important, using an example from your investigation about cellular respiration.

Lab Handout

Lab 2. Photosynthesis: Where Does Photosynthesis Take Place in Plants?

Introduction

Photosynthesis is a chemical process in which green plants produce sugar and oxygen gas (O_2) for themselves. The sugar produced is used by a plant as food to provide energy for other activities. Animals can also eat plants to get sugar for their own energy needs. The O_2 is released into the atmosphere and is also used for other chemical reactions inside plants for producing energy.

The green found in most plants serves an important survival need. The green comes from special organelles found in plants known as *chloroplasts*, which are the location where certain chemical reactions take place that help plants live. Chloroplasts are responsible for giving plants their green color; this color comes from a chemical called *chlorophyll*, which provides green plants with a special chemical ability to absorb energy from light. That energy is used for building sugar molecules during photosynthesis. Photosynthesis requires two chemicals to react: carbon dioxide (CO_2) from the atmosphere and water that can come from the ground (Figure L2.1). The chemical equation for photosynthesis is

Carbon dioxide (CO_2) + water (H_2O) + light energy
→ sugar ($C_6H_{12}O_6$) + oxygen (O_2) + water (H_2O)

This sugar is then used to produce the flowers, leaves, stems, and roots—the *biomass* of the plant. In other words, plants get their building blocks from air! Photosynthesis is the process that plants use to put these building blocks together. The cells not only use the products of photosynthesis to support their own growth but also use those materials to build new cells during reproduction. All cells in a plant need to reproduce at some point, so they will all need those materials. Since cells reproduce everywhere in the plant, it would make sense that photosynthesis would need to happen everywhere, too. But does it?

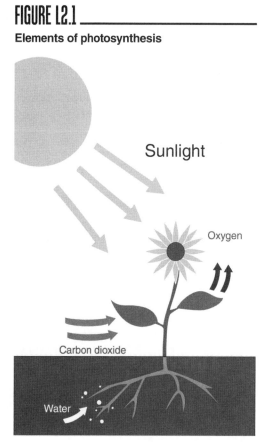

FIGURE L2.1 _____

Elements of photosynthesis

Sunlight

Oxygen

Carbon dioxide

Water

LAB 2

Your Task

Design a scientific investigation to determine if photosynthesis occurs in all of the main parts of a plant. These parts include the flowers, leaves, stems, and roots of a plant.

The guiding question of this investigation is, **Where does photosynthesis take place in plants?**

Materials

You may use any of the following materials during your investigation:

Consumables	Equipment
• Fresh geranium plant	• Scissors
	• O_2 gas sensor
	• CO_2 gas sensor
	• Biochambers or sealed containers with opening for sensors
	• Go!Link adaptor and laptop computer
	• Sanitized indirectly vented chemical-splash goggles
	• Chemical-resistant apron
	• Gloves

Safety Precautions

Follow all normal lab safety rules. In addition, take the following safety precautions:

1. Put on sanitized indirectly vented chemical-splash goggles and laboratory apron and gloves before starting the lab activity.

2. Be careful when using scissors to cut up pieces of the geranium. Also, although geraniums are not very toxic for humans, do not eat any part of the plant.

3. Wash hands with soap and water after completing the lab activity.

Investigation Proposal Required?   ☐ Yes   ☐ No

Getting Started

Figure L2.2 shows how CO_2 and O_2 gas sensors can be inserted into a biochamber. The sensors can then be connected to a laptop to collect data about CO_2 and O_2 gas concentration over periods of time. Ask your teacher for help if you do not understand how to set up the sensors and computer to collect data.

To answer the guiding question, you will need to design and conduct an investigation that explores photosynthesis activity in different parts of a plant. To accomplish this task, you must first determine what type of data you need to collect, how you will collect it, and how you will analyze it. To determine *what type of data you need to collect*, think about the following questions:

- How will you divide the geranium plant up to test for photosynthesis?
- What data will show you that photosynthesis is occurring?
- What type of measurements or observations will you need to record during your investigation?

To determine *how you will collect your data*, think about the following questions:

- What will serve as a control (or comparison) condition?
- What types of treatment conditions will you need to set up and how will you do it?
- How often will you collect data and when will you do it?
- How will you make sure that your data are of high quality (i.e., how will you reduce error)?
- How will you keep track of the data you collect and how will you organize it?

To determine *how you will analyze your data*, think about the following questions:

- How will you determine if there is a difference between the treatment conditions and the control condition?
- What type of calculations will you need to make?
- What type of graph could you create to help make sense of your data?

FIGURE L2.2

CO_2 gas sensor

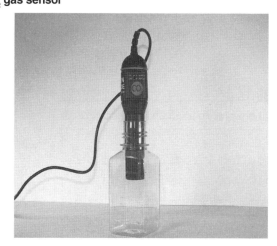

Connections to Crosscutting Concepts, the Nature of Science, and the Nature of Scientific Inquiry

As you work through your investigation, be sure to think about

- how science explores events that happen at different sizes and on different scales,
- how energy and matter flow through living things while being totally conserved,
- how science uses different methods to investigate the natural world, and
- how experiments serve a certain role in science.

LAB 2

Initial Argument

Once your group has finished collecting and analyzing your data, you will need to develop an initial argument. Your argument must include a claim, evidence to support your claim, and a justification of the evidence. The claim is your group's answer to the guiding question. The evidence is an analysis and interpretation of your data. Finally, the justification of the evidence is why your group thinks the evidence matters. The justification of the evidence is important because scientists can use different kinds of evidence to support their claims. Your group will create your initial argument on a whiteboard. Your whiteboard should include all the information shown in Figure L2.3.

FIGURE L2.3

Argument presentation on a whiteboard

The Guiding Question:	
Our Claim:	
Our Evidence:	Our Justification of the Evidence:

Argumentation Session

The argumentation session allows all of the groups to share their arguments. One member of each group will stay at the lab station to share that group's argument, while the other members of the group go to the other lab stations one at a time to listen to and critique the arguments developed by their classmates. This is similar to how scientists present their arguments to other scientists at conferences. If you are responsible for critiquing your classmates' arguments, your goal is to look for mistakes so these mistakes can be fixed and they can make their argument better. The argumentation session is also a good time to think about ways you can make your initial argument better. Scientists must share and critique arguments like this to develop new ideas.

To critique an argument, you might need more information than what is included on the whiteboard. You will therefore need to ask the presenter lots of questions. Here are some good questions to ask:

- What did your group do to collect the data? Why do you think that way is the best way to do it?
- What did your group do to analyze the data? Why did your group decide to analyze it that way?
- What other ways of analyzing and interpreting the data did your group talk about?
- What did your group do to make sure that these calculations are correct?
- Why did your group decide to present your evidence in that way?
- What other claims did your group discuss before you decided on that one? Why did your group abandon those other ideas?
- How sure are you that your group's claim is accurate? What could you do to be more certain?

Once the argumentation session is complete, you will have a chance to meet with your group and revise your original argument. Your group might need to gather more data or design a way to test one or more alternative claims as part of this process. Remember, your goal at this stage of the investigation is to develop the most valid or acceptable answer to the research question!

Report

Once you have completed your research, you will need to prepare an investigation report that consists of three sections that provide answers to the following questions:

1. What question were you trying to answer and why?

2. What did you do during your investigation and why did you conduct your investigation in this way?

3. What is your argument?

Your report should answer these questions in two pages or less. The report must be typed, and any diagrams, figures, or tables should be embedded into the document. Be sure to write in a persuasive style; you are trying to convince others that your claim is acceptable or valid!

Checkout Questions

Lab 2. Photosynthesis: Where Does Photosynthesis Take Place in Plants?

1. Describe the process of photosynthesis and complete the diagram below.

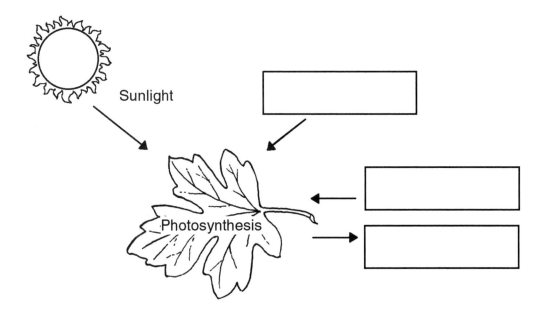

2. Two students were discussing where photosynthesis happens within a plant. One student claimed that photosynthesis only occurs in the leaves of a plant. Another student claimed that photosynthesis occurs in the leaves and flowers of a plant. The table below contains data collected during an investigation. Each container was exposed to the same amount of light for the same amount of time.

Container	Contents	Initial CO_2 level	CO_2 level after 5 hours	CO_2 level after 10 hours
A	3 leaves	900 ppm	700 ppm	500 ppm
B	7 leaves	900 ppm	500 ppm	200 ppm
C	3 leaves and 1 flower	900 ppm	750 ppm	600 ppm
D	2 flowers	900 ppm	950 ppm	1,000 ppm

Use the data above to help support the student you most agree with.

3. The data generated in the scenario for question 2 came from an experiment.

   a. I agree with this statement.

   b. I disagree with this statement.

   Explain your answer.

4. In science, experiments are better than systematic observations.

    a. I agree with this statement.

    b. I disagree with this statement.

Explain your answer, using an example from your investigation about photosynthesis.

5. It is important for scientists to understand the relationship between different quantities and how those relationships change over time. Explain why understanding a relationship between variables is important in science by using an example from your investigation about photosynthesis.

6. It is important for scientists to understand the flow of energy in a system. Explain why this is important, using an example from your investigation about photosynthesis.

Application Labs

LAB 3

Lab Handout

Lab 3. Osmosis: How Does the Concentration of Salt in Water Affect the Rate of Osmosis?

Introduction

In both plants and animals, each cell is surrounded by a membrane. This membrane forms a selective barrier between the cell and its environment (see Figure L3.1—the membrane is the wall in the middle of the figure). Large molecules, such as sugars ($C_6H_{12}O_6$) or fats, and charged molecules, such as sodium ions (Na^+) or chlorine ions (Cl^-), cannot pass through the membrane, but small molecules such as oxygen (O_2) can. Without this barrier, the substances necessary to the life of the cell would diffuse uniformly into the cell's surroundings, and toxic materials from the surroundings would enter the cell. The cell membrane is referred to as *semipermeable* because some particles can naturally cross it while others can-

FIGURE L3.1

A semipermeable membrane

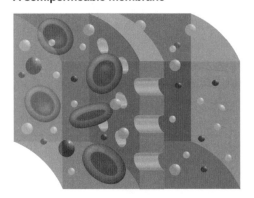

not. This ability to regulate the flow of molecules into and out of the cell keeps the cell's internal environment stable, even though parts of that environment are always shifting.

Chemical particles are constantly in motion. How much they move is related to the amount of energy they contain and how concentrated they are. *Diffusion* is the movement of chemical particles (i.e., atoms, molecules, ions) from an area of high concentration to an area of low concentration. Without any barriers to such movement (like a membrane), chemical particles naturally diffuse in this direction. If a membrane is present, then only particles that can cross it naturally will be able to continue to diffuse normally. To make particles move in the opposite direction (low concentration to high concentration), energy must be added to the particles. *Osmosis* refers specifically to the diffusion of water molecules. In cells, water cannot simply diffuse across the membrane. However, special openings in the membrane allow for easy flow of water molecules so cells can take in or get rid of water when needed.

An *isotonic* solution is a solution that has the same concentration of particles and water as the cell. If blood cells (or other cells) are placed in contact with an isotonic solution, they will neither shrink nor swell. If the solution is *hypertonic*—having a higher concentration of solute (and lower concentration of water) than inside the cell membrane—the cells will lose water and shrink. If the solution is *hypotonic*—having a lower concentration of solute and higher concentration of water molecules—the cells will gain water and swell. Saltwater from the ocean is hypertonic to the cells of the human body since it has more salt in it. Cells, as a result, lose water and shrink (see Figure L3.2). That is why we can't drink water from the ocean—it dehydrates body tissues instead of quenching thirst.

Your Task

Design an experiment to determine how the concentration of salt in water affects the rate of osmosis.

The guiding question of this investigation is, **How does the concentration of salt in water affect the rate of osmosis?**

Materials

You may use any of the following materials during your investigation:

FIGURE L3.2

(a) Red blood cells in saltwater solution and (b) normal red blood cells

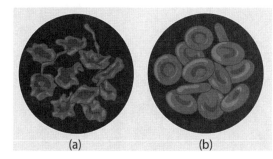

(a)   (b)

Consumables	Equipment
• Salt solutions • Water	• Electronic or triple beam balance • Graduated cylinder and beakers • Dialysis tubing (assume that it behaves just like the membrane of a cell) • Sanitized indirectly vented chemical-splash goggles • Chemical-resistant apron • Gloves

Safety Precautions

Follow all normal lab safety rules. In addition, take the following safety precautions:

1. Put on sanitized indirectly vented chemical-splash goggles and laboratory apron and gloves before starting the lab activity.

2. Immediately wipe up any spilled water to avoid a slip and fall hazard.

3. Wash hands with soap and water after completing the lab activity.

Investigation Proposal Required?   ☐ Yes   ☐ No

Getting Started

You will use models of cells rather than real cells during your experiment. You will use models for two reasons: (1) a model of a cell is much larger than a real cell, which makes the process of data collection much easier; and (2) you can create your cell models in any way you see fit, which makes it easier to control for a wide range of variables during your experiment.

You can construct a model cell by using the dialysis tubing. Dialysis tubing behaves much like a cell membrane. To create a model of a cell, place the dialysis tubing in water

FIGURE L3.3

Tying the dialysis tubing

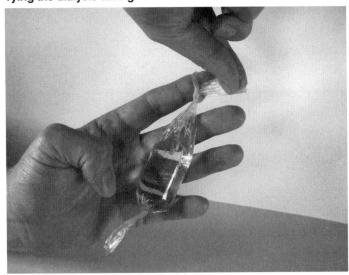

until it is thoroughly soaked. Remove the soaked tubing from the water and tightly twist one end several times and either tie with string or tie a knot in the tubing. You can then fill the model cell with a salt solution or distilled water. Once filled, twist the open end several times and tie it tightly as shown in Figure L3.3. You can then dry the bag and place it into any type of solution you need.

To answer the guiding question, you will need to design and conduct an experiment. To accomplish this task, you must first determine what type of data you need to collect, how you will collect it, and how you will analyze it before you can design your experiment. To determine *what type of data you need to collect*, think about the following questions:

- How will you determine the rate of osmosis?
- What type of measurements or observations will you need to record during your investigation?

To determine *how you will collect your data*, think about the following questions:

- What will serve as a control (or comparison) condition?
- What types of treatment conditions will you need to set up and how will you do it?
- How often will you collect data and when will you do it?
- How will you make sure that your data are of high quality (i.e., how will you reduce error)?
- How will you keep track of the data you collect and how will you organize it?

To determine *how you will analyze your data*, think about the following questions:

- How will you determine if there is a difference between the treatment conditions and the control condition?
- How will you calculate change over time?
- What type of graph could you create to help make sense of your data?

Connections to Crosscutting Concepts, the Nature of Science, and the Nature of Scientific Inquiry

As you work through your investigation, be sure to think about

- why developing and using models is important in science,
- the importance of tracking how matter flows into and out of a system,
- the difference between observations and inferences in science, and
- the nature and role of experiments in science.

Initial Argument

Once your group has finished collecting and analyzing your data, you will need to develop an initial argument. Your argument must include a claim, evidence to support your claim, and a justification of the evidence. The claim is your group's answer to the guiding question. The evidence is an analysis and interpretation of your data. Finally, the justification of the evidence is why your group thinks the evidence matters. The justification of the evidence is important because scientists can use different kinds of evidence to support their claims. Your group will create your initial argument on a whiteboard. Your whiteboard should include all the information shown in Figure L3.4.

FIGURE L3.4 _____

Argument presentation on a whiteboard

The Guiding Question:	
Our Claim:	
Our Evidence:	Our Justification of the Evidence:

Argumentation Session

The argumentation session allows all of the groups to share their arguments. One member of each group will stay at the lab station to share that group's argument, while the other members of the group go to the other lab stations one at a time to listen to and critique the arguments developed by their classmates. This is similar to how scientists present their arguments to other scientists at conferences. If you are responsible for critiquing your classmates' arguments, your goal is to look for mistakes so these mistakes can be fixed and they can make their argument better. The argumentation session is also a good time to think about ways you can make your initial argument better. Scientists must share and critique arguments like this to develop new ideas.

To critique an argument, you might need more information than what is included on the whiteboard. You will therefore need to ask the presenter lots of questions. Here are some good questions to ask:

- What did your group do to collect the data? Why do you think that way is the best way to do it?

- What did your group do to analyze the data? Why did your group decide to analyze it that way?
- What other ways of analyzing and interpreting the data did your group talk about?
- What did your group do to make sure that these calculations are correct?
- Why did your group decide to present your evidence in that way?
- What other claims did your group discuss before you decided on that one? Why did your group abandon those other ideas?
- How sure are you that your group's claim is accurate? What could you do to be more certain?

Once the argumentation session is complete, you will have a chance to meet with your group and revise your initial argument. Your group might need to gather more data or design a way to test one or more alternative claims as part of this process. Remember, your goal at this stage of the investigation is to develop the most valid or acceptable answer to the research question!

Report

Once you have completed your research, you will need to prepare an investigation report that consists of three sections that provide answers to the following questions:

1. What question were you trying to answer and why?

2. What did you do during your investigation and why did you conduct your investigation in this way?

3. What is your argument?

Your report should answer these questions in two pages or less. The report must be typed, and any diagrams, figures, or tables should be embedded into the document. Be sure to write in a persuasive style; you are trying to convince others that your claim is acceptable or valid!

Checkout Questions

Lab 3. Osmosis: How Does the Concentration of Salt in Water Affect the Rate of Osmosis?

1. Describe the process of osmosis.

2. A potato was cut into 10 equal-size cubes, each weighing about 10 grams. The cubes were placed into five different beakers of saltwater, each with a different concentration (%) of salt solution.

The potatoes were allowed to sit in the salt solution for 24 hours and then removed from the beakers, dried, and weighed. The figure below shows the average percent change in mass for the potatoes.

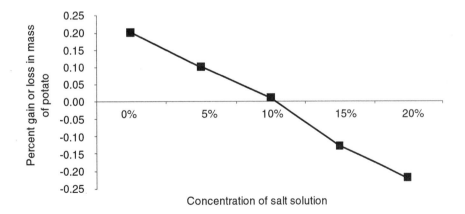

Using what you know about osmosis and the data above, what was the original concentration of salt solution in the potato? Explain your reasoning.

3. The data generated in the scenario for question 2 above came from an experiment.

    a. I agree with this statement.
    b. I disagree with this statement.

   Explain your answer, using an example from your investigation about osmosis.

4. In this investigation we observed osmosis happening.

    a. I agree with this statement.
    b. I disagree with this statement.

   Explain your answer, using an example from your investigation about osmosis.

5. It is important for scientists to develop and use models. Explain why models are important in science by using an example from your investigation about osmosis.

6. It is important for scientists to understand the flow of matter in a system. Explain why this is important, using an example from your investigation about osmosis.

Lab Handout

Lab 4. Cell Structure: What Type of Cell Is on the Unknown Slides?

Introduction

Scientists who study living organisms deal with a lot of different types of life forms, from trees to tadpoles and bacteria to birds. As they investigate how life happens on the planet, they rely on several scientific theories that have developed over time. These theories combine different types of evidence to support a big idea that explains some aspect of life or the natural world. One of the major theories that scientists rely on when studying living things is the *cell theory*. This theory includes three major ideas that have been supported over the years as new life forms continue to be discovered:

1. All living organisms are made up of one or more cells.

2. The cell is the basic unit of life.

3. All new cells come from cells that are already alive.

Just as there are many types of organisms, including plants and animals, there are also many types of cells. However, there are several features found in all cells. The most common features are the presence of DNA and the presence of a *cell membrane*. DNA is a molecule that contains information that cells need to live. The cell membrane is the sheet of molecules that separates the inside of the cell from the rest of the environment. You can think of the cell membrane as a cell's "skin." More complex cells, like those found in animals and plants, have other structures in common, known as *organelles*. Organelles are special structures found inside cells that serve different functions. Those functions include helping the cell get energy, making the materials it needs to continue growing, and storing the information (like DNA) to make new cells. The organelles present in a cell will also influence what activities that cell can perform.

Plant and animal cells have many organelles in common, including the nucleus, the endoplasmic reticulum, Golgi bodies, ribosomes, the cell membrane, and mitochondria (see Figure L4.1, p. 38). Some organelles found in plant cells, however, are not found in animal cells, and vice versa. For example, animal cells have centrioles (which help organize cell division in animal cells), but plant cells do not. Plant cells have an extra layer surrounding them called a cell wall. Cell walls are stiff membranes that sit outside of the cell membrane and help keep plant cells in a specific shape. The differences in types of organelles can be used to distinguish between cells that come from a plant and cells that come from an animal. However, not all organelles can be seen using microscopes we use in school.

LAB 4

Animal cell diagram

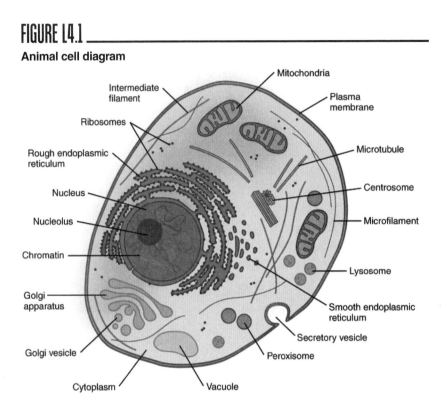

Your Task

Using what you learn from observations of several slides of cells using a microscope, with some slides labeled as plant cells and others as animal cells, determine what types of cells are on the slides labeled as "unknown."

The guiding question of this investigation is, **What type of cell is on the unknown slides?**

Materials

You may use any of the following materials during your investigation:

- Known slide A (plant cells)
- Known slide B (plant cells)
- Known slide C (animal cells)
- Known slide D (animal cells)
- Unknown slide E
- Unknown slide F
- Unknown slide G

- Unknown slide H
- Compound light microscope
- Slide wipes
- Sanitized indirectly vented chemical-splash goggles
- Chemical-resistant apron
- Gloves

Safety Precautions

Follow all normal lab safety rules. In addition, take the following safety precautions:

1. Put on sanitized indirectly vented chemical-splash goggles and laboratory apron and gloves before starting the lab activity.

2. Handle all glassware with care to avoid breakage. Sharp edges can cut skin!

3. Follow all safety rules that apply when working with electrical equipment, and use only GFCI-protected electrical receptacles.

4. Wash hands with soap and water after completing the lab activity.

Investigation Proposal Required? ☐ Yes ☐ No

Getting Started

To determine the difference between a plant cell and animal cell, you and your group will need to explore what cell structures you can see with a compound light microscope. To answer the guiding question, you must first determine what type of data you need to collect, how you will collect it, and how you will analyze it. To determine *what type of data you need to collect*, think about the following questions:

- What type of measurements or observations will you need to make during your investigation?
- How will you quantify any differences or similarities you observe in the different cells?

To determine *how you will collect your data*, think about the following questions:

- How will you make sure that your data are of high quality (i.e., how will you reduce error)?
- How will you keep track of the data you collect and how will you organize it?

To determine *how you will analyze your data*, think about the following question:

- How will you define the different categories of cells (e.g., what makes a plant cell a plant cell, what makes an animal cell an animal cell)?

Connections to Crosscutting Concepts, the Nature of Science, the Nature of Scientific Inquiry

As you work through your investigation, be sure to think about

- how scientists look for patterns across different living things,

LAB 4

- how the structure of an organelle or cell is related to the function it performs,
- the difference between observations and inferences, and
- how science knowledge changes over time as new evidence is discovered and technology is created.

Initial Argument

Once your group has finished collecting and analyzing your data, you will need to develop an initial argument. Your argument must include a claim, evidence to support your claim, and a justification of the evidence. The claim is your group's answer to the guiding question. The evidence is an analysis and interpretation of your data. Finally, the justification of the evidence is why your group thinks the evidence matters. The justification of the evidence is important because scientists can use different kinds of evidence to support their claims. Your group will create your initial argument on a whiteboard. Your whiteboard should include all the information shown in Figure L4.2.

FIGURE L4.2

Argument presentation on a whiteboard

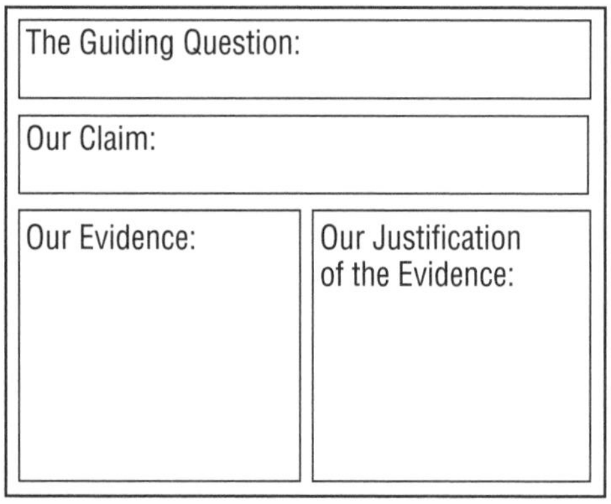

The Guiding Question:	
Our Claim:	
Our Evidence:	Our Justification of the Evidence:

Argumentation Session

The argumentation session allows all of the groups to share their arguments. One member of each group will stay at the lab station to share that group's argument, while the other members of the group go to the other lab stations one at a time to listen to and critique the arguments developed by their classmates. This is similar to how scientists present their arguments to other scientists at conferences. If you are responsible for critiquing your classmates' arguments, your goal is to look for mistakes so these mistakes can be fixed and they can make their argument better. The argumentation session is also a good time to think about ways you can make your initial argument better. Scientists must share and critique arguments like this to develop new ideas.

To critique an argument, you might need more information than what is included on the whiteboard. You will therefore need to ask the presenter lots of questions. Here are some good questions to ask:

- What did your group do to collect the data? Why do you think that way is the best way to do it?
- What did your group do to analyze the data? Why did your group decide to analyze it that way?
- What other ways of analyzing and interpreting the data did your group talk about?
- Why did your group decide to present your evidence in that way?

- What other claims did your group discuss before you decided on that one? Why did your group abandon those other ideas?
- How sure are you that your group's claim is accurate? What could you do to be more certain?

Once the argumentation session is complete, you will have a chance to meet with your group and revise your original argument. Your group might need to gather more data or design a way to test one or more alternative claims as part of this process. Remember, your goal at this stage of the investigation is to develop the most valid or acceptable answer to the research question!

Report

Once you have completed your research, you will need to prepare an investigation report that consists of three sections that provide answers to the following questions:

1. What question were you trying to answer and why?

2. What did you do during your investigation and why did you conduct your investigation in this way?

3. What is your argument?

Your report should answer these questions in two pages or less. This report must be typed, and any diagrams, figures, or tables should be embedded into the document. Be sure to write in a persuasive style; you are trying to convince others that your claim is acceptable or valid!

Checkout Questions

Lab 4. Cell Structure: What Type of Cell Is on the Unknown Slides?

1. Describe the characteristics that plant cells and animal cells have in common.

2. Describe the features that allow you to distinguish between plant cells and animal cells.

3. In science, it is not possible to make an inference without first observing.

   a. I agree with this statement.
   b. I disagree with this statement.

   Explain your answer, using an example from your investigation about cell structure.

4. Once a scientific idea is developed, it does not change.

    a. I agree with this statement.

    b. I disagree with this statement.

    Explain your answer, using an example from your investigation about cell structure.

5. It is important for scientists to look for and identify patterns in nature. Explain why identifying patterns is useful in science by using an example from your investigation about cell structure.

6. It is important for scientists to understand the relationship between the structure of an organism and its function. Explain why this is important, using an example from your investigation about cell structure.

LAB 5

Lab 5. Temperature and Photosynthesis: How Does Temperature Affect the Rate of Photosynthesis in Plants?

Introduction

All the colors that our eyes can see can be found in plants that live in the world. They include deep, bright reds to rich blues and purples. Interestingly, one common color that most plants share is green. This common characteristic is not simply due to chance. The green found in most plants actually serves an important survival need. The green comes from special organelles found in plants known as *chloroplasts*. Chloroplasts are the location where certain chemical reactions take place that help plants live. Chloroplasts are responsible for giving plants their green color; this color comes from a chemical called chlorophyll, which provides green plants with a special chemical ability to absorb energy from light. Green plants have the ability to produce their own supply of sugar through the process of *photosynthesis*.

FIGURE L5.1 _____

Elements of photosynthesis

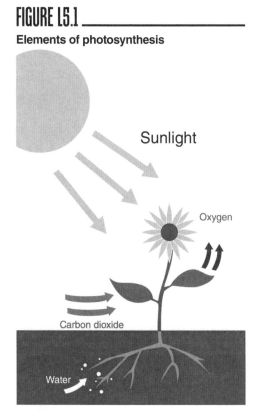

Photosynthesis is a complex chemical process in which green plants produce sugar and oxygen gas (O_2) for themselves. The sugar produced is used by a plant as food to provide energy for other activities. Animals can also eat plants to get sugar for their own energy needs. The O_2 is released into the atmosphere (see Figure L5.1) and is used for other chemical reactions inside plants for producing energy. Animals also use O_2 to produce energy. This common process for producing energy in living things is known as *respiration*, which refers to a series of chemical reactions that use O_2 to break bonds in the sugar molecules, releasing the energy stored there. The products of respiration include carbon dioxide gas (CO_2), which is released into the atmosphere and can then be used in photosynthesis.

Photosynthesis requires two chemicals to react: CO_2 from the atmosphere and water that can come from the ground. However, photosynthesis also requires light energy to make sugar and O_2. The chemical equation for photosynthesis is

$$\text{Carbon dioxide } (CO_2) + \text{water } (H_2O) + \text{light energy}$$
$$\rightarrow \text{sugar } (C_6H_{12}O_6) + \text{oxygen } (O_2) + \text{water } (H_2O)$$

This sugar is then used to produce the flowers, leaves, stems, and roots—the *biomass* of the plant. In other words, plants get their building blocks from air!

Plants rely on gases in the air and sunlight to create the food and energy they need. The amount of sunlight in a plant's environment will influence how much photosynthesis can occur. The amount of water present (dry desert vs. moist rainforest) also influences plants' activities and affects their growth. Obviously, plant growth and survival depend in part on conditions in the environment. A common example of this is the difference between the seasons. Most plants do not show a lot of activity during the winter, although they remain alive. In the spring, they become colorful and grow a lot. That growth continues over the summer, with some plants producing fruits and vegetables, and then their growth slows down during the fall. We also know that the temperature in the environment is very different across the seasons. So if overall plant growth and behavior are affected by seasonal changes in temperature, how do different temperatures affect the process of photosynthesis?

Your Task

Design a scientific investigation to determine how temperature influences the rate of photosynthesis in plants.

The guiding question of this investigation is, **How does temperature affect the rate of photosynthesis in plants?**

Materials

You may use any of the following materials during your investigation:

Consumables	Equipment
• Fresh geranium plant	• O_2 gas sensor
	• CO_2 gas sensor
	• Temperature probe
	• Go!Link adaptor and laptop computer
	• Biochamber or sealed container with opening for sensors
	• Hot plate or incubator
	• Ice packs or ice bath
	• Sanitized indirectly vented chemical-splash goggles
	• Chemical-resistant apron
	• Gloves

Safety Precautions

Follow all normal lab safety rules. In addition, take the following safety precautions:

1. Put on sanitized indirectly vented chemical-splash goggles and laboratory apron and gloves before starting the lab activity.

LAB 5

2. Be careful handling hot plates and incubators set at high temperatures because they may be hot enough to burn you.

3. Wash hands with soap and water after completing the lab activity.

Investigation Proposal Required? ☐ Yes ☐ No

Getting Started

FIGURE L5.2

CO_2 gas sensor

Figure L5.2 shows how CO_2 and O_2 gas sensors can be inserted into a biochamber. The sensors can then be connected to a laptop to collect data about CO_2 and O_2 gas concentration over periods of time. Ask your teacher for help if you do not understand how to set up the sensors and computer to collect data.

To answer the guiding question, you will need to design and conduct an investigation that explores rates of photosynthesis in different temperatures. To accomplish this task, you must determine what type of data you need to collect, how you will collect it, and how you will analyze it. To determine *what type of data you need to collect*, think about the following questions:

• How will you test the effect of temperature on the photosynthesis rate?

• How will you change the temperature inside the chamber?

• What type of measurements or observations will you need to record during your investigation?

To determine *how you will collect your data*, think about the following questions:

• What will serve as a control (or comparison) condition?

• What types of treatment conditions will you need to set up and how will you do it?

• How often will you collect data and when will you do it?

• How will you make sure that your data are of high quality (i.e., how will you reduce error)?

• How will you keep track of the data you collect and how will you organize it?

To determine *how you will analyze your data*, think about the following questions:

- How will you determine if there is a difference between the treatment conditions and the control condition?
- What type of calculations will you need to make?
- What type of graph could you create to help make sense of your data?

Connections to Crosscutting Concepts, the Nature of Science, and the Nature of Scientific Inquiry

As you work through your investigation, be sure to think about

- how scientists try to figure out cause-and-effect relationships that explain why something happens,
- how energy and matter flow through living things while being totally conserved,
- the different roles theories and laws play in science, and
- the difference between data collected and evidence created in an investigation.

Initial Argument

Once your group has finished collecting and analyzing your data, you will need to develop an initial argument. Your argument must include a claim, evidence to support your claim, and a justification of the evidence. The claim is your group's answer to the guiding question. The evidence is an analysis and interpretation of your data. Finally, the justification of the evidence is why your group thinks the evidence matters. The justification of the evidence is important because scientists can use different kinds of evidence to support their claims. Your group will create your initial argument on a whiteboard. Your whiteboard should include all the information shown in Figure L5.3.

FIGURE L5.3

Argument presentation on a whiteboard

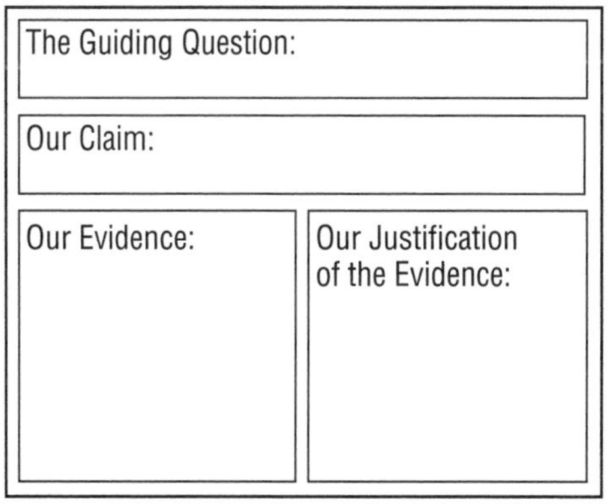

The Guiding Question:	
Our Claim:	
Our Evidence:	Our Justification of the Evidence:

Argumentation Session

The argumentation session allows all of the groups to share their arguments. One member of each group will stay at the lab station to share that group's argument, while the other members of the group go to the other lab stations one at a time to listen to and critique the arguments developed by their classmates. This is similar to how scientists present their arguments to other scientists at conferences. If you are responsible for critiquing your classmates' arguments, your goal is to look for mistakes so these mistakes can be fixed and they can make their argument better. The argumentation session is also a good time to think about ways you can make your initial argument better. Scientists must share and critique arguments like this to develop new ideas.

LAB 5

To critique an argument, you might need more information than what is included on the whiteboard. You will therefore need to ask the presenter lots of questions. Here are some good questions to ask:

- What did your group do to collect the data? Why do you think that way is the best way to do it?
- What did your group do to analyze the data? Why did your group decide to analyze it that way?
- What other ways of analyzing and interpreting the data did your group talk about?
- What did your group do to make sure that these calculations are correct?
- Why did your group decide to present your evidence in that way?
- What other claims did your group discuss before you decided on that one? Why did your group abandon those other ideas?
- How sure are you that your group's claim is accurate? What could you do to be more certain?

Once the argumentation session is complete, you will have a chance to meet with your group and revise your original argument. Your group might need to gather more data or design a way to test one or more alternative claims as part of this process. Remember, your goal at this stage of the investigation is to develop the most valid or acceptable answer to the research question!

Report

Once you have completed your research, you will need to prepare an investigation report that consists of three sections that provide answers to the following questions:

1. What question were you trying to answer and why?
2. What did you do during your investigation and why did you conduct your investigation in this way?
3. What is your argument?

Your report should answer these questions in two pages or less. This report must be typed, and any diagrams, figures, or tables should be embedded into the document. Be sure to write in a persuasive style; you are trying to convince others that your claim is acceptable or valid!

Checkout Questions

Lab 5. Temperature and Photosynthesis: How Does Temperature Affect the Rate of Photosynthesis in Plants?

Use the following information to answer questions 1 and 2.

Andre owns a tree farm where he grows trees and then sells them to people to plant in their yard. Andre can't sell his trees until they grow to 8 feet tall, which takes about six months. The faster the trees grow, the sooner he can sell them and make more money. Andre wanted to investigate how he could help his trees grow faster without spending money on extra fertilizer or water for the plants. He had the idea of growing a sample of trees in a greenhouse where he kept the temperature warm, about 85°F (29.4°C), and another sample in a greenhouse where he kept the temperature cooler, about 65°F (18.3°C). Andre watered all the plants the same and made sure both groups got the same amount of sunlight. After just four months the trees in the warm greenhouse had all reached 8 feet tall, but the trees in the cooler greenhouse only grew to 5 feet.

1. For a plant to grow, what must be valid about the rate of photosynthesis compared with the rate of cellular respiration? Explain your reasoning.

2. Use what you know about photosynthesis and temperature to explain why the trees in the warm greenhouse grew taller than the ones in the cool greenhouse.

3. A theory and a law serve the same purpose in science.

   a. I agree with this statement.
   b. I disagree with this statement.

Explain your answer, using an example from your investigation about temperature and photosynthesis.

4. In science, it is not possible to make an inference without first observing.

    a. I agree with this statement.
    b. I disagree with this statement.

Explain your answer, using an example from your investigation about temperature and photosynthesis.

5. Scientists try to identify the effect that certain factors can cause in a system. Explain why identifying cause-and-effect relationships is useful in science, using an example from your investigation about temperature and photosynthesis.

6. It is important for scientists to understand the relationship between energy and matter and how they move through a system. Explain why this is important, using an example from your investigation about temperature and photosynthesis.

Lab Handout

Lab 6. Energy in Food: Which Type of Nut Is Best for a New Energy Bar?

Introduction

All living things must take in nutrients to survive. In plants, the process of photosynthesis converts light energy from the sun, along with carbon dioxide and water to generate sugar molecules that the plant then uses for chemical energy to complete cellular processes that are required for the plant to live and grow. Animals (including humans), however, must consume food to obtain the nutrients and chemical energy they need to complete the cellular processes that allow them to live and grow. People get energy to live by consuming a variety of food types, such as carbohydrates, proteins, and fats. Each of these food types has different properties and provides our bodies with different amounts of energy. Plants and animals break down nutrients to release chemical energy through a process called *cellular respiration*. During cellular respiration, large molecules containing carbon, hydrogen, and oxygen atoms are broken down into smaller molecules in a process similar to the combustion (or burning) of wood in a fire.

The amount of chemical energy that is stored in a food source is measured in *food calories*. A food calorie is also called a *kilocalorie* and is defined as the amount of energy needed to heat 1 kilogram of water by 1 degree Celsius. The chemical energy stored in food is released through the processes of digestion and cellular respiration so that our bodies can then use the energy for important tasks such as moving our muscles. Different food types contain different amounts of energy that our bodies can use; in general, fats provide the most energy at about 9 food calories per gram. Carbohydrates and proteins both provide about 4 food calories per gram.

Many foods contain a mix of carbohydrates, protein, and fats; this is true of nuts (peanuts, pecans, almonds, etc.). The nutrients found in different foods are shown on the *nutrition facts* label that is included on most food items sold in the United States (see Figure L6.1 [p. 52] for an example of this label). These labels also include information about the types of vitamins found in foods, such as vitamin A or vitamin C, or minerals found in food, such as calcium or iron. Vitamins and minerals are important for certain cellular processes, but they do not provide any calories or energy for our bodies.

Each day our bodies need a minimum number of calories so that we have enough energy to complete the tasks that keep us alive, such as maintaining our core body temperature, breathing, and keeping our heart beating and circulating blood and oxygen throughout our body. The amount of calories an individual needs to complete the most basic functions of life is called the *resting metabolic rate*. Resting metabolic rates differ from person

LAB 6

Example of a nutrition facts label

Nutrition Facts
Serving Size 2/3 cup (55g)
Servings Per Container About 8

Amount Per Serving

Calories 230   Calories from Fat 40

	% Daily Value*
Total Fat 8g	**12%**
Saturated Fat 1g	**5%**
Trans Fat 0g	
Cholesterol 0mg	**0%**
Sodium 160mg	**7%**
Total Carbohydrate 37g	**12%**
Dietary Fiber 4g	**16%**
Sugars 1g	
Protein 3g	

Vitamin A	10%
Vitamin C	8%
Calcium	20%
Iron	45%

* Percent Daily Values are based on a 2,000 calorie diet. Your daily value may be higher or lower depending on your calorie needs.

	Calories:	2,000	2,500
Total Fat	Less than	65g	80g
Sat Fat	Less than	20g	25g
Cholesterol	Less than	300mg	300mg
Sodium	Less than	2,400mg	2,400mg
Total Carbohydrate		300g	375g
Dietary Fiber		25g	30g

to person based on gender, height, weight, and many other factors. People need additional energy so they can walk, talk, or do other everyday tasks, and people who exercise or play sports need even more energy so that they can complete the tasks associated with those activities such as running, jumping, or lifting heavy objects. To get all the energy we need on a daily basis, we must eat enough food to provide our bodies with all the calories we need. On average, an adult should consume about 2,400 food calories a day. In addition to eating three meals a day, sometimes it is necessary to have a snack to take in enough calories for the day. Many athletes, for example, must eat "energy bars" before they exercise or even during long exercise sessions to make sure they get the extra calories they need.

Your Task

Collect data to help you determine which type of nut would be the best to use for a new energy bar. To do this, you will need to measure the amount of energy in a variety of nuts and consider several other factors, such as cost, size of bar, and/or amount of calories that are necessary to complete different exercises.

The guiding question of this investigation is, **Which type of nut is best for a new energy bar?**

Materials

You may use any of the following materials during your investigation:

Consumables	Equipment
• Nuts (peanut, cashew, pecan, almond, walnut)	• Calorimeter
• Matches	• Combustion stand
	• Thermometer or temperature sensor
	• Graduated cylinder (100 ml)
	• Timer
	• Electronic or triple beam balance
	• Sanitized indirectly vented chemical-splash goggles
	• Chemical-resistant apron
	• Gloves
	• Lab 6 Reference Sheet

Safety Precautions

Even though there are food items used during this lab investigation, they should be treated as chemicals and not eaten. In addition, take the following safety precautions:

1. Notify the teacher immediately if you have an allergy to nuts.

2. Put on sanitized indirectly vented chemical-splash goggles and laboratory apron and gloves before starting the lab activity.

3. Never put consumables in your mouth.

4. Use caution when working with open flames while burning the nuts and using matches. Open flames can burn skin, and combustibles and flammables must be kept away from the open flame. If you have long hair, tie it back behind your head.

5. Handle all glassware with care to avoid breakage. Sharp glass can cut skin!

6. Wash your hands with soap and water after completing the lab activity.

Investigation Proposal Required?   ☐ Yes   ☐ No

Getting Started

During your investigation you will need to determine the amount of energy in each type of nut. To do this, you can use a calorimeter to measure the amount of heat energy released by burning each type of nut (see Figure L6.2). Measuring the temperature change in a sample of water will provide information related to the amount of energy stored in the nut.

To answer the guiding question, you will also need to consider other factors such as the cost of different types of nuts and how many calories it takes to complete various exercises. To accomplish this task, you must first determine what type of data you need to collect, how you will collect it, and how you will analyze it. To determine *what type of data you need to collect*, think about the following questions:

• What information will tell you which nut(s) have the most energy?

• How will the calorimeter and thermometer (or temperature sensor) help you measure the energy in the nuts?

• What type of measurements or observations will you need to record during your investigation?

To determine *how you will collect your data*, think about the following questions:

• What will serve as a control (or comparison) condition?

• What types of treatment conditions will you need to set up and how will you do it?

• How often will you collect data and when will you do it?

• How will you make sure that your data are of high quality (i.e., how will you reduce error)?

FIGURE L6.2 _____

A simple calorimeter that can be used to measure the amount of energy in a nut

- How will you keep track of the data you collect and how will you organize it?

To determine *how you will analyze your data*, think about the following questions:

- How will you determine if there is a difference between the treatment conditions and the control condition?
- What type of calculations will you need to make?
- What type of graph could you create to help make sense of your data?
- How many calories should your energy bar have?
- How much will your energy bar cost?

Connections to Crosscutting Concepts, the Nature of Science, and the Nature of Scientific Inquiry

As you work through your investigation, be sure to think about

- how energy and matter flow through living things while being totally conserved,
- how an object's shape or structure determines many of its properties and functions,
- the difference between data collected and evidence created in an investigation, and
- the role of imagination and creativity when solving problems in science.

Initial Argument

Once your group has finished collecting and analyzing your data, you will need to develop an initial argument. Your argument must include a claim, evidence to support your claim, and a justification of the evidence. The claim is your group's answer to the guiding question. The evidence is an analysis and interpretation of your data. Finally, the justification of the evidence is why your group thinks the evidence matters. The justification of the evidence is important because scientists can use different kinds of evidence to support their claims. Your group will create your initial argument on a whiteboard. Your whiteboard should include all the information shown in Figure L6.3.

FIGURE L6.3

Argument presentation on a whiteboard

The Guiding Question:	
Our Claim:	
Our Evidence:	Our Justification of the Evidence:

Argumentation Session

The argumentation session allows all of the groups to share their arguments. One member of each group will stay at the lab station to share that group's argument, while the other members of the group go to the other lab stations one at a time to listen to and critique the arguments developed by their classmates. This is similar to how scientists present their arguments to other scientists at conferences. If you are responsible for critiquing your classmates' arguments, your goal is to look for mistakes so these mistakes can be fixed and they can make their argument better. The argumentation session is also a good time to think about ways you can make your initial argument better. Scientists must share and critique arguments like this to develop new ideas.

To critique an argument, you might need more information than what is included on the whiteboard. You will therefore need to ask the presenter lots of questions. Here are some good questions to ask:

- What did your group do to collect the data? Why do you think that way is the best way to do it?
- What did your group do to analyze the data? Why did your group decide to analyze it that way?
- What other ways of analyzing and interpreting the data did your group talk about?
- What did your group do to make sure that these calculations are correct?
- Why did your group decide to present your evidence in that way?
- What other claims did your group discuss before you decided on that one? Why did your group abandon those other ideas?
- How sure are you that your group's claim is accurate? What could you do to be more certain?

Once the argumentation session is complete, you will have a chance to meet with your group and revise your original argument. Your group might need to gather more data or design a way to test one or more alternative claims as part of this process. Remember, your goal at this stage of the investigation is to develop the most valid or acceptable answer to the research question!

Report

Once you have completed your research, you will need to prepare an investigation report that consists of three sections that provide answers to the following questions:

1. What question were you trying to answer and why?
2. What did you do during your investigation and why did you conduct your investigation in this way?
3. What is your argument?

Your report should answer these questions in two pages or less. This report must be typed, and any diagrams, figures, or tables should be embedded into the document. Be sure to write in a persuasive style; you are trying to convince others that your claim is acceptable or valid!

LAB 6

Lab 6 Reference Sheet

Costs and Exercise Calories

As you decide which nut will be best for your energy bar, be sure to consider how much of each type of nut would need to go into your bar and how much that will cost. The following table gives cost per pound for each type of nut used in this lab investigation.

Type of nut	Cost per pound
Peanut	$2.00
Cashew	$5.92
Pecan	$7.73
Almond	$8.88
Walnut	$10.43

Your energy bar should be made so that it stays together and maintains its shape. Many energy bars use honey or maple syrup as a binding agent to help the pieces stay together. The honey or syrup also adds flavor to the energy bar. It takes about 15 grams of honey or syrup per 100 grams of nuts for the energy bar to stay intact. The following table gives calories and cost for these binding agents.

Binding agent	Calories (per 100 g)	Cost (per 100 g)
Honey	304	$0.44
Maple syrup	260	$0.60

For your energy bar to be useful, it should supply enough calories that will provide energy to complete at least 30 minutes of exercise activity. The following table lists various activities and how much energy is used during 1 hour of that activity.

Activity	Calories (per hour)
Running (6 mph)	600
Walking (2 mph)	200
Biking (12 mph)	480
Swimming	420
Dancing	220
Aerobic exercise (high impact)	500
Martial arts	700
Jumping rope	850
Lifting weights	365

Checkout Questions

Lab 6. Energy in Food: Which Type of Nut Is Best for a New Energy Bar?

1. Steven made the claim that the energy we get from the food we eat can be traced all the way back to the Sun. Use what you know about energy transfer, photosynthesis, and cellular respiration to support his claim.

2. The energy we get from nutrients like carbohydrates and fats is a result of breaking chemical bonds and releasing the stored chemical potential energy. One gram of carbohydrate provides 4 food calories, but one gram of fat provides 9 food calories. Use what you know about energy and molecules to describe why fats provide more energy (per gram) than carbohydrates.

3. In the energy bar investigation, you collected data that you used to develop evidence.

   a. I agree with this statement.

   b. I disagree with this statement.

   Explain your answer, using an example from your investigation about energy in food.

4. Scientists do not need to be creative or have a good imagination to excel in science.

   a. I agree with this statement.
   b. I disagree with this statement.

   Explain your answer, using an example from your investigation about energy in food.

5. An important goal in science is to develop an understanding of how matter and energy move through complex systems. Explain why understanding the flow of matter and energy is important, using an example from your investigation about energy in food.

6. Scientists often investigate how the structure or composition of something is related to its function. Explain how composition and function are related, using an example from your investigation about energy in food.

Lab Handout

Lab 7. Respiratory and Cardiovascular Systems: How Do Activity and Physical Factors Relate to Respiratory and Cardiovascular Fitness?

Introduction

All bodies, including those of humans, animals, plants, and bacteria, are made of cells. Some of these organisms (humans, animals, and plants) have their cells arranged into groups that all help the organisms to live. In humans and animals, scientists have identified several different body systems. Body systems are very organized groups of cells that serve specific functions that allow organisms to live. Although these systems are basically large groups of cells, those cells are organized into other levels. Pieces at each level work together to help pieces at the next level work. Body systems are made up of smaller units called organs, like the heart and lungs, which help the body take in and transport the nutrients it needs. Organs are made of up tissues, like muscle fibers in the heart, which help the organs function. For example, muscle tissue in the heart helps it to beat so it can transport materials through blood. Tissues are made up of cells, such as muscle cells that make up muscle fibers. Muscle cells shift proteins around to make muscle fibers grow and shrink so the whole muscle contracts, causing the heart to beat, which helps the rest of the systems work.

The human body is made up of several systems. The functions these systems perform include breaking down food we eat (digestive), directing the activities of the body (nervous), and helping us fight off disease (immune). The respiratory system (Figure L7.1) includes the lungs and trachea that breathe in air, containing oxygen (O_2) for use in the body, and breathe out waste gases, like carbon dioxide (CO_2). The cardiovascular system (Figure L7.2; also called the circulatory system) uses blood, arteries, veins, and

FIGURE L7.1
Respiratory system

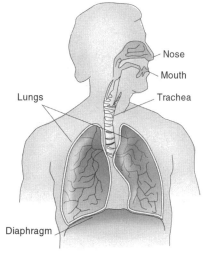

FIGURE L7.2
Cardiovascular system

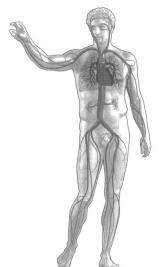

LAB 7

the heart to deliver materials such as O_2 from your lungs. It also removes waste, like CO_2, from cells; regulates body temperature; and helps the body fight off disease.

As you can see, the function of these two body systems is connected. They both work on moving gases into and out of the body. Cells in our body use O_2 to help make energy through a chemical process called cellular respiration. The same chemical process gives off CO_2, which is transported through blood from cells to the lungs to be breathed out. The gas exchange function of these two body systems helps all the cells in the body. This is because most cells must use O_2 to create energy.

Cellular respiration uses O_2 to release energy from molecules in food we eat. A waste product from this process is CO_2, the same gas that we breathe out. So as our body takes in and puts out these gases, so do the cells in our body. Indeed, that is why we breathe more when we exercise. We know that humans use respiration to produce energy because when a human breathes, the air that he or she inhales contains about 21% O_2 and less than 1% CO_2; however, when he or she exhales, the air contains about 15% O_2 and 5% CO_2. Our cells need more O_2 to make more energy to support the increased physical activity.

The functions of these systems are connected to the amount of work the body does. So personal activity levels, like being an athlete or doing regular exercise, can influence how well these systems work. Also, the structure of these systems is influenced by physical factors such as height or frame size. Larger people may have more cells than smaller people, and having more cells means higher energy demands. Physical activity can also increase the number of cells in the body. Gender (boy or girl) can influence the function of different systems. Environmental factors can also affect the functions of the body; some locations have higher or lower amounts of O_2 available in the air. With less O_2 in the air, the body has to breathe more to get the O_2 it needs for activity. The function of these body systems is tied to the energy needs of the body.

Your Task

Design an investigation to explore the relationship between fellow students' cardiovascular and respiratory fitness and different activity and physical factors.

The guiding question of this investigation is, **How do activity and physical factors relate to respiratory and cardiovascular fitness?**

Materials

You may use any of the following materials during your investigation:

- Elastic straps
- Transmitter belt
- Exercise heart rate monitor
- Timer
- Go!Link adaptor and laptop computer

- Stool (1–2 feet high)
- Cardiovascular fitness protocol
- Cardiovascular fitness table
- Lab 7 Reference Sheet (1 per student)

Safety Precautions

Follow all normal lab safety rules. In addition, take the following safety precautions:

1. Notify your teacher if you have heart or other health conditions.

2. The heart rate monitor contains electrical connections covered by metal plates that are safe to touch your skin. However, only use the heart rate monitor equipment as directed. Any playing or non-approved use of the equipment could potentially hurt you or your classmates.

Investigation Proposal Required?   ☐ Yes   ☐ No

Getting Started

You will be given a Lab 7 Reference Sheet that includes a cardiovascular fitness protocol and a cardiovascular fitness table; these resources will guide the data collection process. After reviewing this sheet, you will need to determine what type of data you need to collect, how you will collect the data, and how you will analyze the data. To determine *what type of data you need to collect*, think about the following questions:

- What personal or physical factors could influence respiratory and cardiovascular fitness?

- What type of measurements or observations will you need to record during your investigation?

To determine *how you will collect your data*, think about the following questions:

- What types of treatment conditions will you need to set up and how will you do it?

- During the experiment, when will you collect data and how often will you collect it?

- How will you make sure that your data are of high quality (i.e., how will you reduce error)?

- How will you keep track of the data you collect and how will you organize it?

To determine *how you will analyze your data*, think about the following questions:

- How will you determine overall cardiovascular fitness?

- How will you compare subgroups?

- What type of calculations will you need to make?

- What type of graph could you create to help make sense of your data?

Connections to Crosscutting Concepts, the Nature of Science, and the Nature of Scientific Inquiry

As you work through your investigation, be sure to think about

- how scientists study systems by creating models of their structure and function,
- how the structure of living things affects the way they function,
- the way science is influenced by the society in which it takes place, and
- the specific role that experiments play in science.

Initial Argument

Once your group has finished collecting and analyzing your data, you will need to develop an initial argument. Your argument must include a claim, evidence to support your claim, and a justification of the evidence. The claim is your group's answer to the guiding question. The evidence is an analysis and interpretation of your data. Finally, the justification of the evidence is why your group thinks the evidence matters. The justification of the evidence is important because scientists can use different kinds of evidence to support their claims. Your group will create your initial argument on a whiteboard. Your whiteboard should include all the information shown in Figure L7.3.

FIGURE L7.3 _____

Argument presentation on a whiteboard

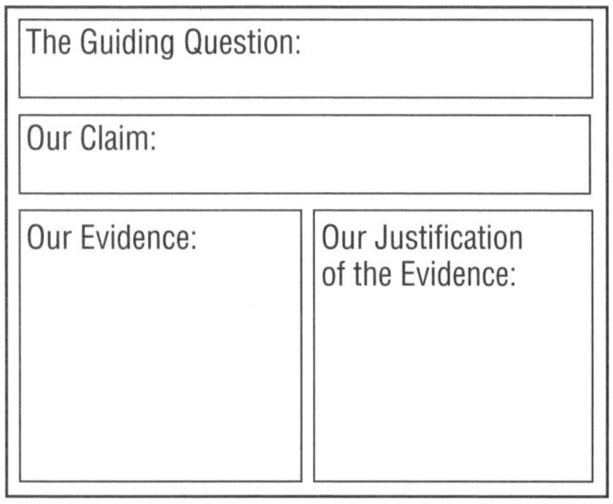

The Guiding Question:	
Our Claim:	
Our Evidence:	Our Justification of the Evidence:

Argumentation Session

The argumentation session allows all of the groups to share their arguments. One member of each group will stay at the lab station to share that group's argument, while the other members of the group go to the other lab stations one at a time to listen to and critique the arguments developed by their classmates. This is similar to how scientists present their arguments to other scientists at conferences. If you are responsible for critiquing your classmates' arguments, your goal is to look for mistakes so these mistakes can be fixed and they can make their argument better. The argumentation session is also a good time to think about ways you can make your initial argument better. Scientists must share and critique arguments like this to develop new ideas.

To critique an argument, you might need more information than what is included on the whiteboard. You will therefore need to ask the presenter lots of questions. Here are some good questions to ask:

- What did your group do to collect the data? Why do you think that way is the best way to do it?

- What did your group do to analyze the data? Why did your group decide to analyze it that way?

- What other ways of analyzing and interpreting the data did your group talk about?

- What did your group do to make sure that these calculations are correct?

- Why did your group decide to present your evidence in that way?

- What other claims did your group discuss before you decided on that one? Why did your group abandon those other ideas?

- How sure are you that your group's claim is accurate? What could you do to be more certain?

Once the argumentation session is complete, you will have a chance to meet with your group and revise your original argument. Your group might need to gather more data or design a way to test one or more alternative claims as part of this process. Remember, your goal at this stage of the investigation is to develop the most valid or acceptable answer to the research question!

Report

Once you have completed your research, you will need to prepare an investigation report that consists of three sections that provide answers to the following questions:

1. What question were you trying to answer and why?

2. What did you do during your investigation and why did you conduct your investigation in this way?

3. What is your argument?

Your report should answer these questions in two pages or less. This report must be typed, and any diagrams, figures, or tables should be embedded into the document. Be sure to write in a persuasive style; you are trying to convince others that your claim is acceptable or valid!

LAB 7

Lab 7 Reference Sheet

Cardiovascular Fitness Test Protocol and Tables

Cardiovascular Fitness Test Protocol

To determine the cardiovascular health of a person, you will need to measure reclining heart rate, exercise heart rate (step test), and exercise recovery time.[1]

Reclining Heart Rate

1. Have the subject lie on a clean surface or table. Begin collecting heart rate data. While the subject is lying on the table, count the number of breaths he or she takes in one minute.

2. Record the subject's heart rate after two minutes in the cardiovascular fitness table.

3. Assign fitness points to the subject based on Table R7.1 and record the value in the cardiovascular fitness table. Also, record the subject's respiratory rate in the cardiovascular fitness table.

TABLE R7.1 _____

Reclining heart rate fitness points

Beats/min.	Fitness points
50–60	6
61–70	5
71–80	4
81–90	3
91–100	2
101–110	1
> 110	0

---

1 This protocol has been adapted from a similar protocol available from Vernier, Inc., for use with their heart rate monitors.

Step Test

4. Before performing the step test, have the subject stand still for 30 seconds. Record the subject's heart rate at 30 seconds as the subject's pre-exercise heart rate.

5. Perform a step test using the following procedure:

   a. Place the right foot on the top step of the stool.

   b. Place the left foot completely on the top step of the stool next to the right foot.

   c. Place the right foot back on the floor.

   d. Place the left foot completely on the floor next to the right foot.

   e. Repeat the cycle as quickly as possible for 30 seconds.

6. Record the heart rate in the subject's cardiovascular fitness table. **DO NOT STOP DATA COLLECTION!** Start timer and quickly move to step 7. While working on steps 7 and 8 for heart rate, measure the subject's *respiratory* rate in 30 seconds and multiply that number by 2. (You can start step 7 and then begin measuring respiratory rate.)

Exercise Recovery Time

7. Have the subject remain standing and keep relatively still. Monitor the heart rate readings and stop timing when the readings return to the pre-exercise heart rate value recorded in step 4. Record the recovery time in the cardiovascular fitness table.

8. Locate the subject's recovery time in Table R7.2 and record the corresponding fitness point value in the cardiovascular fitness table. Also, record the subject's respiratory rate when he or she has reached recovery heart rate.

TABLE R7.2

Recovery time fitness points

Time (sec.)	Fitness points
0–30	6
31–60	5
61–90	4
91–120	3
121–150	2
151–180	1
>180	0

9. Subtract the subject's pre-exercise heart rate from his or her heart rate after five stepping cycles of exercise. Record this heart rate increase in the endurance row of the cardiovascular fitness table. Do the same subtraction for the respiratory rate.

10. Locate the row corresponding to the pre-exercise heart rate in Table R7.3 and use the heart rate increase value to determine endurance fitness points. Record the subject's endurance fitness points in the cardiovascular fitness table.

TABLE R7.3

Endurance fitness points

Reclining heart rate (beats/min.)	Heart rate increase after exercise (beats/min.)				
	0–10	11–20	21–30	31–40	41+
50–60	6	5	4	3	2
61–70	5	4	3	2	1
71–80	4	3	2	1	0
81–90	3	2	1	0	0
91–100	2	1	0	0	0
101–110	1	0	0	0	0
>110	0	0	0	0	0

Cardiovascular Fitness Table

Subject number: _____   Gender: _____   Age: _____

Frame size: _____   Height: _____   Factor: _____

Condition	Value	Fitness points	Respiratory rate
Reclining heart rate	beats/min.		breaths/min.
Pre-exercise heart rate	beats/min.		breaths/min.
Step test	beats/min.		breaths/min.
Exercise recovery time	seconds		
Endurance	beats/min.		breaths/min.
		Total:	

Lab 7. Respiratory and Cardiovascular Systems: How Do Activity and Physical Factors Relate to Respiratory and Cardiovascular Fitness?

There are many systems in the body that must work together to sustain life. Systems such as the respiratory system, circulatory system, and digestive system are involved in making sure our bodies have the energy it needs to function.

1. Use what you know about energy transfer and each of these systems to describe how they work together to help our body function.

2. Jeremy made the claim in science class that most body systems work together, but the nervous system is the only one that operates in isolation. Do you agree with Jeremy? Explain your reasoning.

3. Scientists should not allow their society or culture to influence their work.

   a. I agree with this statement.

   b. I disagree with this statement.

   Explain your answer, using an example from your investigation about the respiratory and cardiovascular systems.

4. Investigations in medical science often involve people, so experiments are not used.

   a. I agree with this statement.

   b. I disagree with this statement.

   Explain your answer, using an example from your investigation about the respiratory and cardiovascular systems.

5. Scientists often generate models when they are working with complex systems or events. Explain why using models in science is helpful, using an example from your investigation about the respiratory and cardiovascular systems.

6. Scientists often investigate how the structure of something is related to its function. Explain how structure and function are related, using an example from your investigation about the respiratory and cardiovascular systems.

LAB 8

Lab 8. Memory and Stimuli: How Does the Way Information Is Presented Affect Working Memory?

Introduction

The human body is made up of several systems that work together. These systems perform certain activities necessary for living. For example, the respiratory system (which includes the lungs) allows us to take in oxygen and get rid of carbon dioxide, and the digestive system (which includes the stomach) allows us to break down and use the food we eat. The *nervous system* is responsible for taking in information and directing the actions of other parts of the body. This system includes nerves and the brain. The nerves carry information around the body and the brain. The brain sends and receives information, controlling many different activities at the same time.

The brain gets information about the world around the body through *senses*. There are five senses: hearing, seeing, tasting, smelling, and touching. Each of these senses uses a special organ that is filled with lots of nerves. Those nerves send information to the brain that tells us what is going on in the world. The *eyes* are the organ for seeing. The eyes take in light waves to send information to the brain, and the brain uses that information to understand the world around us. But what does the brain do with that information?

Our senses respond to *stimuli*, which is the plural of *stimulus*. A stimulus is a thing or event that evokes a specific functional reaction in an organ or tissue. All things our eyes see, including the words on this page, are stimuli. Our senses take in information about the stimuli and send it to the brain by chemical and electrical signals. The brain reads that information and acts on it. The brain reads and uses the information in several ways:

- *Short-term memory* keeps information for only a few seconds. Your short-term memory handles the most basic information, like the light level in a room.

- *Working memory* keeps information for just a little longer, allowing us to organize it and make sense of it. You are using your working memory as you read this sentence.

- *Long-term memory* keeps lots of information for long periods of times, up to years and decades.

As information is processed, the brain makes connections across it and organizes it based on patterns. Using those patterns, the brain is able to take in more information faster and retain it longer. Information that is presented in a pattern makes it easier for our brains to make sense of the world.

Your Task

Design an investigation to see how much information people can store in their working memory. Your goal is to explore how the amount of information and the order in which it is presented affects what people can remember.

The guiding question of this investigation is, **How does the way information is presented affect working memory?**

Materials

You may use any of the following materials during your investigation:

- Set of cards numbered 1 through 9
- Set of memory letter cards
- Paper
- Timer

Safety Precautions

Follow all normal lab safety rules.

Investigation Proposal Required? ☐ Yes ☐ No

Getting Started

You will be given a set of cards to see how many numbers people can remember. For this test, you lay out one numbered card and give a person 20 seconds to memorize it. Cover the card with a piece of paper, and then have the person tell you the number. Lay out another card beside the first one. Give the person another 20 seconds to memorize the two numbers. Cover the card and then have the person tell you the numbers in the correct order. Keep adding more cards until the person cannot tell you the numbers in the correct order.

You will also be given a set of cards with the same group of letters on them but in different orders. For this test, lay out Card 1 in the set and give a person 30 seconds to memorize it. Take the card away and have the person tell you the letters that were on the card. Lay out Card 2 in the set and give the person 30 seconds to memorize it. Take the card away and have the person tell you the letters that were on the card. Do the same thing using Card 3.

To answer the guiding question, you will need to determine what type of data you need to collect, how you will collect the data, and how you will analyze the data. To determine *what type of data you need to collect*, think about the following questions:

- What kind of information can you get from the person telling you about what's on the card?

- What type of measurements or observations will you need to record during your investigation?

To determine *how you will collect the data*, think about the following questions:

- What types of conditions will you need to set up and how will you do it?
- During the experiment, when will you collect data and how often will you collect it?
- How will you make sure that your data are of high quality (i.e., how will you reduce error)?
- How will you keep track of the data you collect and how will you organize it?

To determine *how you will analyze the data*, think about the following questions:

- How will you connect information about people's memories to the types of cards you used?
- How will you compare subgroups?
- What type of calculations will you need to make?
- What type of graph could you create to help make sense of your data?

Connections to Crosscutting Concepts, the Nature of Science, and the Nature of Scientific Inquiry

As you work through your investigation, be sure to think about

- how scientists look for patterns in the world,
- how the structure of living things affects the way they function,
- the way science is influenced by the society in which it takes place, and
- the role of imagination and creativity when solving problems in science.

Initial Argument

Once your group has finished collecting and analyzing your data, you will need to develop an initial argument. Your argument must include a claim, evidence to support your claim, and a justification of the evidence. The claim is your group's answer to the guiding question. The evidence is an analysis and interpretation of your data. Finally, the justification of the evidence is why your group thinks the evidence matters. The justification of the evidence is important because scientists can use different kinds of evidence to support their claims. Your group will create your initial argument on a whiteboard. Your whiteboard should include all the information shown in Figure L8.1.

Argumentation Session

The argumentation session allows all of the groups to share their arguments. One member of each group will stay at the lab station to share that group's argument, while the other members of the group go to the other lab stations one at a time to listen to and critique the arguments developed by their classmates. This is similar to how scientists present their arguments to other scientists at conferences. If you are responsible for critiquing your classmates' arguments, your goal is to look for mistakes so these mistakes can be fixed and they can make their argument better. The argumentation session is also a good time to think about ways you can make your initial argument better. Scientists must share and critique arguments like this to develop new ideas.

FIGURE L8.1

Argument presentation on a whiteboard

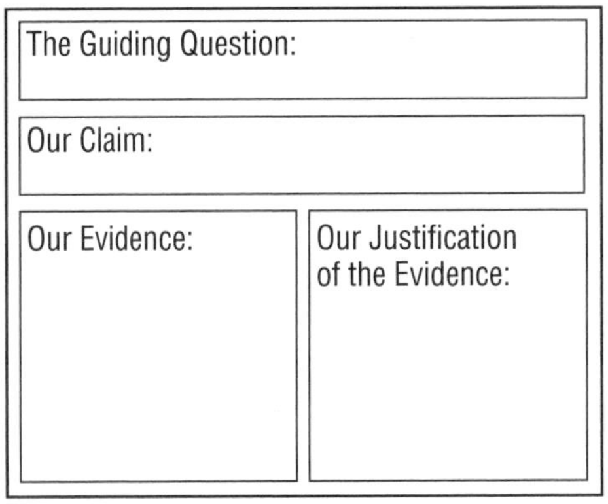

The Guiding Question:	
Our Claim:	
Our Evidence:	Our Justification of the Evidence:

To critique an argument, you might need more information than what is included on the whiteboard. You will therefore need to ask the presenter lots of questions. Here are some good questions to ask:

- What did your group do to collect the data? Why do you think that way is the best way to do it?
- What did your group do to analyze the data? Why did your group decide to analyze it that way?
- What other ways of analyzing and interpreting the data did your group talk about?
- What did your group do to make sure that these calculations are correct?
- Why did your group decide to present your evidence in that way?
- What other claims did your group discuss before you decided on that one? Why did your group abandon those other ideas?
- How sure are you that your group's claim is accurate? What could you do to be more certain?

Once the argumentation session is complete, you will have a chance to meet with your group and revise your original argument. Your group might need to gather more data or design a way to test one or more alternative claims as part of this process. Remember, your goal at this stage of the investigation is to develop the most valid or acceptable answer to the research question!

Report

Once you have completed your research, you will need to prepare an investigation report that consists of three sections that provide answers to the following questions:

LAB 8

1. What question were you trying to answer and why?

2. What did you do during your investigation and why did you conduct your investigation in this way?

3. What is your argument?

Your report should answer these questions in two pages or less. This report must be typed, and any diagrams, figures, or tables should be embedded into the document. Be sure to write in a persuasive style; you are trying to convince others that your claim is acceptable or valid!

Checkout Questions

Lab 8. Memory and Stimuli: How Does the Way Information Is Presented Affect Working Memory?

1. Phone numbers are organized in a 3-3-4 pattern, for example: 555-867-5309. Use what you know about working memory to explain why organizing phone numbers in this way is more helpful that listing all 10 numbers in a row.

2. The two images below were made with six small sticks. If a person has only has one second to look at the images and then has to redraw them, which image do you think would be easier to remember? Explain your reasoning.

Image A

Image B

3. Science is influenced by the society in which it takes place.

   a. I agree with this statement.

   b. I disagree with this statement.

Explain your answer, using an example from your investigation about memory and stimuli.

4. Imagination and creativity have no place in scientific investigations.

    a. I agree with this statement.
    b. I disagree with this statement.

Explain your answer, using an example from your investigation about memory and stimuli.

5. Scientists often look for patterns when they are investigating the world. Explain why identifying patterns in the world is helpful, using an example from your investigation about memory and stimuli.

6. Scientists often investigate how the structure of something is related to its function. Explain how structure and function are related in terms of body systems, using an example from your investigation about memory and stimuli.

SECTION 3

Life Sciences
Core Idea 2

Ecosystems: Interactions, Energy, and Dynamics

Introduction Labs

LAB 9

Lab Handout

Lab 9. Population Growth: What Factors Limit the Size of a Population of Yeast?

Introduction

All populations of living things change in size over time. The human population is no different. A population is a group of individuals that belong to the same species and live in the same region at the same time. In 1800, the total number of people on Earth was about 1 billion. Now the world population is 7 billion. This observation has caused many scientists to ask questions such as "What caused this dramatic increase in the human population?" and "How long will the human population continue to grow?"

Human population growth has been a "hot topic" of discussion in the popular media and among ecologists (scientists who study how organisms interact with each other and the environment). To better understand human population growth on Earth, it is necessary to study population dynamics, or the changing size, density, and range of a population. Population dynamics is an area of life science that focuses on these changes. Scientists studying *population dynamics* investigate how different factors in living organisms and the environment shape those changes. However, humans are not the only living things studied through population dynamics. Factors related to all living organisms can include what they eat and drink, the amount of space they need, and how they interact with other organisms. Many of these factors are considered *biotic factors*, because they involve living things in an ecosystem. Factors related to the environment can include how much space is available and the weather and climate patterns in a particular area. Many of these factors are considered *abiotic factors*, because they involve nonliving pieces of an ecosystem.

By investigating these biotic and abiotic factors, scientists determine important relationships that help populations grow or reduce the number of organisms in an area. Scientists study the population dynamics of many different organisms to help us understand more about what affects human populations. Also, scientists study how human populations change other populations of organisms. Knowing about these relationships helps scientists and policy makers make decisions about how to manage natural resources and organisms in many different types of ecosystems.

Your Task

Design an investigation on how the size of a population of yeast changes over time in response to different factors such as *amount of food*, *amount of space*, and *the initial size of the population*. Yeasts are single-celled organisms in the Fungi kingdom. One species of yeast called *Saccharomyces cerevisiae* (Figure L9.1) has been used in baking and alcoholic beverages for thousands of years. Scientists have also used it to gather information about how cells function because it reproduces quickly. In fact, *S. cerevisiae* and many other species of yeast can produce a new generation every two hours. Therefore, a population of yeast could potentially increase in size very quickly if something did not prevent the size of the population from growing over time.

The guiding question of this investigation is, **What factors limit the size of a population of yeast?**

FIGURE L9.1 _____

Microscopic image of yeast (*Saccharomyces cerevisiae*)

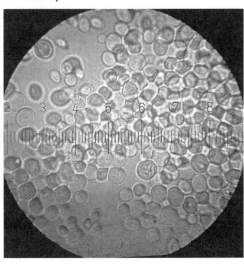

Materials

You may use any of the following materials during your investigation:

Consumables	Equipment
• Yeast culture • Sugar solution • Iodine solution • Distilled water	• Graduated pipette (1 ml) • Test tubes • Test tube rack • Simple compound light microscope • Microscope slides • Cover slips • Calculator • Sanitized indirectly vented chemical-splash goggles • Chemical-resistant apron • Gloves

Safety Precautions

Follow all normal lab safety rules. In addition, take the following safety precautions:

1. If you have any allergies to yeast, be sure to discuss this with your teacher immediately!

2. Put on sanitized indirectly vented chemical-splash goggles and laboratory apron and gloves before starting the lab activity.

3. Handle all glassware with care to avoid breakage. Sharp glass edges can cut skin!

LAB 9

4. Review the important information on chemicals on the safety data sheet, and use caution when handling chemicals.

5. Follow all safety rules that apply when working with electrical equipment, and use only GFCI-protected electrical receptacles.

6. Wash hands with soap and water after completing the lab activity.

Investigation Proposal Required?   ☐ Yes   ☐ No

Getting Started

Brainstorm with your group one possible factor that you think will limit the size of the yeast population. While designing your investigation, consider this variable while writing your hypotheses (e.g., population size is limited by the amount of food available, space available, and other factors). Be sure to include a control condition in your investigation, which will include environmental conditions that you will not change across different treatment conditions. A control condition usually represents the "normal" environment for the organism you are studying. You will need to track the size of a population of yeast for the next 72 hours.

Listed below are some important tips for working with yeast.

Setting up a population of yeast:

1. Use a graduated pipette to transfer 1 ml of the yeast from the class culture to a standard test tube. Measure carefully. In this case, more is not better.

2. Add two drops of iodine to yeast in the test tube. Be sure to drop the iodine into the culture, not on the side of the test tube. (The iodine will help you to see the cells under the microscope.)

3. Add water to the test tube (this can range from 1 ml to 5 ml depending on how much space you want to give the yeast population).

4. Add sugar solution to the test tube (this can range from 1 ml to 3 ml depending on how much food you want to give the yeast population).

Counting the number of yeast in your test tube:

1. Because yeast cells tend to settle out of solution, you will need to stir the yeast in your test tube so that the cells are evenly distributed. This must be done gently to avoid foaming the culture.

2. Use the 1 ml pipette to transfer 0.1 ml (a single drop) from the test tube to the graduated microscope slide.

3. Carefully lower a cover slip onto the drop to make a wet mount slide. Observe the slide under low power and identify the yeast cells.

4. Count the number of yeast cells in three different fields of view under high power. Select those fields of view from different areas of the slide.

5. Add the total number of cells you counted in all three squares and find the average number of cells per field of view.

To answer the guiding question, you must first determine what type of data you need to collect, how you will collect it, and how you will analyze it. To determine *what type of data you need to collect*, think about the following question:

- What type of measurements or observations will you need to record during your investigation?

To determine *how you will collect your data*, think about the following questions:

- What will serve as a control (or comparison) condition?
- What types of treatment conditions will you need to set up and how will you do it?
- During the experiment, when will you collect data and how often will you collect it?
- How will you make sure that your data are of high quality (i.e., how will you reduce error)?
- How will you keep track of the data you collect and how will you organize it?

To determine *how you will analyze your data*, think about the following questions:

- How will you determine if there is a difference between the treatment conditions and the control condition?
- How will you calculate change over time?
- What type of graph could you create to help make sense of your data?

Connections to Crosscutting Concepts, the Nature of Science, and the Nature of Scientific Inquiry

As you work through your investigation, be sure to think about

- how scientists try to identify patterns in nature to better understand it,
- how living things go through periods of stability followed by periods of change,
- the difference between observations and inferences in science, and
- the role of experiments in science

LAB 9

Initial Argument

Once your group has finished collecting and analyzing your data, you will need to develop an initial argument. Your argument must include a claim, evidence to support your claim, and a justification of the evidence. The claim is your group's answer to the guiding question. The evidence is an analysis and interpretation of your data. Finally, the justification of the evidence is why your group thinks the evidence matters. The justification of the evidence is important because scientists can use different kinds of evidence to support their claims. Your group will create your initial argument on a whiteboard. Your whiteboard should include all the information shown in Figure L9.2.

FIGURE L9.2
Argument presentation on a whiteboard

The Guiding Question:	
Our Claim:	
Our Evidence:	Our Justification of the Evidence:

Argumentation Session

The argumentation session allows all of the groups to share their arguments. One member of each group will stay at the lab station to share that group's argument, while the other members of the group go to the other lab stations one at a time to listen to and critique the arguments developed by their classmates. This is similar to how scientists present their arguments to other scientists at conferences. If you are responsible for critiquing your classmates' arguments, your goal is to look for mistakes so these mistakes can be fixed and they can make their argument better. The argumentation session is also a good time to think about ways you can make your initial argument better. Scientists must share and critique arguments like this to develop new ideas.

To critique an argument, you might need more information than what is included on the whiteboard. You will therefore need to ask the presenter lots of questions. Here are some good questions to ask:

- What did your group do to collect the data? Why do you think that way is the best way to do it?
- What did your group do to analyze the data? Why did your group decide to analyze it that way?
- What other ways of analyzing and interpreting the data did your group talk about?
- What did your group do to make sure that these calculations are correct?
- Why did your group decide to present your evidence in that way?
- What other claims did your group discuss before you decided on that one? Why did your group abandon those other ideas?
- How sure are you that your group's claim is accurate? What could you do to be more certain?

Once the argumentation session is complete, you will have a chance to meet with your group and revise your original argument. Your group might need to gather more data or design a way to test one or more alternative claims as part of this process. Remember, your goal at this stage of the investigation is to develop the most valid or acceptable answer to the research question!

Report

Once you have completed your research, you will need to prepare an investigation report that consists of three sections that provide answers to the following questions:

1. What question were you trying to answer and why?

2. What did you do during your investigation and why did you conduct your investigation in this way?

3. What is your argument?

Your report should answer these questions in two pages or less. The report must be typed, and any diagrams, figures, or tables should be embedded into the document. Be sure to write in a persuasive style; you are trying to convince others that your claim is acceptable or valid!

Checkout Questions

Lab 9. Population Growth: What Factors Limit the Size of a Population of Yeast?

1. Describe how the amount of food or amount of space available might influence the size of a population of organisms.

2. In a laboratory, yeast can be grown in a very controlled setting without many external influences on their population. In nature, however, conditions are constantly changing. Using what you know about population dynamics, describe what might happen to a population of organisms if the climate where they live changes.

3. *Observation* and *inference* are two words that mean the same thing.

   a. I agree with this statement.
   b. I disagree with this statement.

   Explain your answer, using an example from your investigation about population growth.

4. If scientists want to be certain about an idea, they must conduct an experiment to test it.

    a. I agree with this statement.

    b. I disagree with this statement.

    Explain your answer, using an example from your investigation about population growth.

5. Scientists often look for patterns when they are investigating the world. Explain why identifying patterns in the world is helpful, using an example from your investigation about population growth.

6. Understanding how living things go through periods of stability followed by periods of change is important for scientists. Explain why understanding the relationship between stability and change is important, using an example from your investigation about population growth.

LAB 10

Lab 10. Predator-Prey Relationships: How Is the Size of a Predator Population Related to the Size of a Prey Population?

Introduction

John Donne wrote, "No man is an island." The same is true for any individual plant or animal. Individuals are always part of a larger group of organisms from the same species, called a *population*. In some respects, populations act like individual organisms. They require space and nutrients. They have daily and seasonal cycles, including when they sleep and eat. They grow and die. The size of any population is in large part determined by a balance between several factors; some of these factors are obvious and some are not. Ultimately the biological success of a population is measured by its size over time. The factors that affect how well organisms grow and survive are the factors that determine population size.

A population of one kind of organism also interacts with other populations in the same area where it lives. Several populations interacting with each other form a *community*. Organisms from different populations but from the same community relate in several different ways. Some organisms from different populations help each other survive through providing resources for one another, which is known as *mutualism*. An example of this relationship would be bees and flowering plants. The bees find food when the flowers bloom on the plant, and that food helps the bee population survive. When bees visit multiple flowers, they also carry pollen from one flower to another. This movement of pollen helps the flowering plants to reproduce, so the flower population continues to survive.

Another kind of relationship found between populations is known as *predation*. Predation involves organisms from one population using organisms from another population as food. The organism that is used as food is called *prey*, and the organism that eats the other organism is called a *predator*. There are many examples of predator-prey relationships. When you pick vegetables from a garden, you are a predator and the plant is prey. Seagulls and bears are predators of several kinds of fish that are their prey. Predator-prey relationships are very common in different communities of organisms.

Your Task

Using a computer simulation, investigate how a population of a predator (wolves) and a population of its prey (sheep) interact with each other and the local environment over time.

The guiding question of this investigation is, **How is the size of a predator population related to the size of a prey population?**

Materials

You will use an online simulation called *Wolf Sheep Predation* to conduct your investigation. You can find the simulation in the module library of the NetLogo program on your lab computer: *http://ccl.northwestern.edu/netlogo/models/WolfSheepPredation.*

Safety Precautions

Follow all normal lab safety rules.

Investigation Proposal Required?   ☐ Yes   ☐ No

Getting Started

The *Wolf Sheep Predation* simulation allows you to explore how populations of predators and prey interact with each other over time. This simulation is designed to follow the rules of nature so you can use it to see what happens to the size of a population when you change different environmental factors, such as size of initial population and availability of food for the prey species.

In the simulation, wolves and sheep wander randomly around the landscape. When a wolf bumps into a sheep, the wolf eats the sheep. Each step costs the wolves energy, and they must eat sheep to replenish their energy; when they run out of energy, they die. Each wolf or sheep reproduces at a constant rate. You can also choose to include grass in addition to wolves and sheep in the simulation. If you add the grass, the sheep must eat grass to maintain their energy; when they run out of energy, they die. Once grass is eaten, it will only regrow after a fixed amount of time.

You can change a wide range of factors in the simulation:

- INITIAL-NUMBER-SHEEP: the initial size of the sheep population
- INITIAL-NUMBER-WOLVES: the initial size of the wolf population
- WOLF-GAIN-FROM-FOOD: the amount of energy a wolf gets for every sheep eaten
- SHEEP-REPRODUCE: how often sheep reproduce
- WOLF-REPRODUCE: how often wolves reproduce

To answer the guiding question, you must determine what type of data you need to collect, how you will collect it, and how you will analyze it. To determine *what type of data you need to collect*, think about the following questions:

- How will you determine if the composition of the sheep and wolf populations changes over time?

- What will serve as your dependent variable?
- What type of measurements or observations will you need to record during your investigation?

To determine *how you will collect your data*, think about the following questions:

- What will serve as a control condition?
- What types of treatment conditions will you need to set up and how will you do it?
- How many trials will you need to conduct?
- How long will you need to run the simulation during each trial?
- How often will you collect data and when will you do it?
- How will you keep track of the data you collect and how will you organize it?

To determine *how you will analyze your data*, think about the following questions:

- How will you determine if there is a difference between the different treatment conditions and the control condition?
- What type of calculations will you need to make?
- What type of graph could you create to help make sense of your data?

Connections to Crosscutting Concepts, the Nature of Science, and the Nature of Scientific Inquiry

As you work through your investigation, be sure to think about

- the importance of understanding cause-and-effect relationships for natural phenomena,
- the use of models to study systems,
- the way scientific knowledge can change over time, and
- the different types of methods that scientists use to answer questions.

Initial Argument

Once your group has finished collecting and analyzing your data, you will need to develop an initial argument. Your argument must include a claim, evidence to support your claim, and a justification of the evidence. The claim is your group's answer to the guiding question. The evidence is an analysis and interpretation of your data. Finally, the justification of the evidence is why your group thinks the evidence matters. The justification of the evidence is important because scientists can use different kinds of evidence to support their claims. Your group will create your initial argument on a whiteboard. Your whiteboard should include all the information shown in Figure L10.1.

Argumentation Session

The argumentation session allows all of the groups to share their arguments. One member of each group will stay at the lab station to share that group's argument, while the other members of the group go to the other lab stations one at a time to listen to and critique the arguments developed by their classmates. This is similar to how scientists present their arguments to other scientists at conferences. If you are responsible for critiquing your classmates' arguments, your goal is to look for mistakes so these mistakes can be fixed and they can make their argument better. The argumentation session is also a good time to think about ways you can make your initial argument better. Scientists must share and critique arguments like this to develop new ideas.

FIGURE L10.1

Argument presentation on a whiteboard

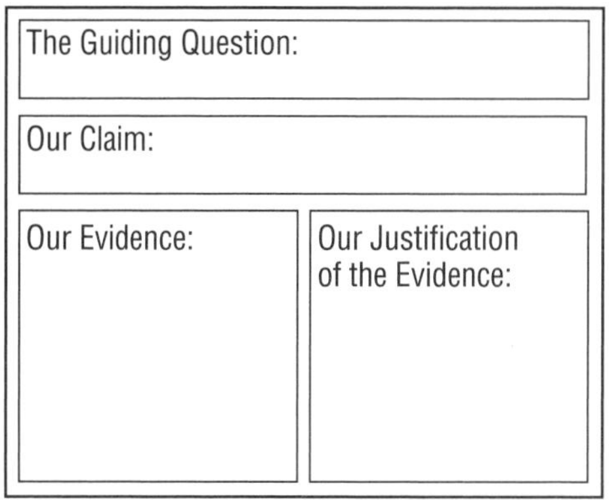

The Guiding Question:	
Our Claim:	
Our Evidence:	Our Justification of the Evidence:

To critique an argument, you might need more information than what is included on the whiteboard. You will therefore need to ask the presenter lots of questions. Here are some good questions to ask:

- What did your group do to collect the data? Why do you think that way is the best way to do it?

- What did your group do to analyze the data? Why did your group decide to analyze it that way?

- What other ways of analyzing and interpreting the data did your group talk about?

- Why did your group decide to present your evidence in that way?

- What other claims did your group discuss before you decided on that one? Why did your group abandon those other ideas?

- How sure are you that your group's claim is accurate? What could you do to be more certain?

Once the argumentation session is complete, you will have a chance to meet with your group and revise your original argument. Your group might need to gather more data or design a way to test one or more alternative claims as part of this process. Remember, your goal at this stage of the investigation is to develop the most valid or acceptable answer to the research question!

Report

Once you have completed your research, you will need to prepare an investigation report that consists of three sections that provide answers to the following questions:

LAB 10

1. What question were you trying to answer and why?

2. What did you do during your investigation and why did you conduct your investigation in this way?

3. What is your argument?

Your report should answer these questions in two pages or less. The report must be typed, and any diagrams, figures, or tables should be embedded into the document. Be sure to write in a persuasive style; you are trying to convince others that your claim is acceptable or valid!

National Science Teachers Association

Checkout Questions

Lab 10. Predator-Prey Relationships: How Is the Size of a Predator Population Related to the Size of a Prey Population?

1. In some ecosystems there may be multiple predators. Using what you know about predator-prey relationships, describe how it is possible for multiple predators to exist in the same habitat.

2. In an ecosystem, the size of the predator population is related to the size of the prey population. Describe what would happen if the predator population reproduced faster than the prey population.

3. Scientific knowledge changes so quickly that it should be considered unstable.

   a. I agree with this statement.

   b. I disagree with this statement.

   Explain your answer, using an example from your investigation about predator-prey relationships.

4. It is important for scientists to use a variety of methods to learn about the natural world.

    a. I agree with this statement.

    b. I disagree with this statement.

Explain your answer, using an example from your investigation about predator-prey relationships.

5. Scientists often study potential cause-and-effect relationships when they investigate the natural world. Explain why it is important to understand causes and effects, using an example from your investigation about predator-prey relationships.

6. Scientists develop models to help them understand the natural world. Sometimes the models scientists develop are similar to the computer model that you used in the predator-prey investigation. Explain why models are helpful, using an example from your investigation about predator-prey relationships.

Application Labs

LAB 11

Lab Handout

Lab 11. Food Webs and Ecosystems: Which Member of an Ecosystem Would Affect the Food Web the Most If Removed?

Introduction

An *ecosystem* includes all the living and nonliving pieces of a particular area of the planet. Living things in an ecosystem must eat other living things in the ecosystem to get the energy they need to survive. The only organisms that do not have to eat other organisms for their energy are called *producers*. Producers are organisms that create their own food by harvesting energy from other sources, such as the Sun. Plants are the most common type of producers found in an ecosystem. If an organism is not a producer in an ecosystem, then it is considered a *consumer*. Consumers are organisms that have to eat other living things to get the energy they need to survive. Some consumers will eat only the plants in an ecosystem, some consumers will eat only other consumers, and still other consumers will eat both the plants and other consumers.

Different organisms have different energy needs, which will influence what food they eat. In any ecosystem, there can be multiple producers and types of consumers. One way that scientists try to understand these relationships in an ecosystem is through designing *food webs*. A food web is a diagram that models the feeding relationships in an ecosystem. It can also be considered the combination of all the unique *food chains* present in an ecosystem. Food chains are models that represent the eating relationship among a group of organisms present in an ecosystem. There can be many food chains present in a single ecosystem. One species of organism can be involved in multiple food chains. Food webs help show all the individual food chains operating in an ecosystem and how they overlap.

Figure L11.1 provides an example of a food web. Notice how each organism has line arrows pointed into them and other line arrows coming out from them. A line with an arrow coming out of an organism indicates what that organism eats; in contrast, a line with an arrow pointing into an organism indicates that the organism is eaten by the organism at the other end of the line. Also notice how one type of organism in the food web can be a food source for several other organisms in the same ecosystem.

FIGURE L11.1 _____

Example of a food web diagram, showing the eating relationships in an ecosystem

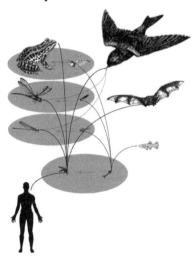

By understanding the food web of a certain ecosystem, scientists can also understand the impact human activity can have on that ecosystem. There are many situations in which humans try to remove a certain type of organism from an ecosystem, often for reasons involving public health or managing resources. Humans can add chemicals to an ecosystem that can get rid of certain plants or insects from an ecosystem. They can also hunt larger organisms that may be a higher-level organism in an ecosystem's food web. However, eliminating one type of organism from an ecosystem will have an impact on other organisms in that system.

Your Task

Explore the different roles of organisms in a specific ecosystem. A town has to decide which organism it should remove from its local ecosystem, which includes a swampy marsh. Many residents are worried about the mosquitoes that heavily populate the marsh. Others are concerned with the growth in algae and other weedlike plants in the marsh. Still other residents believe that the ducks in the marsh are a problem and should be hunted. Removing any one of these organisms, or others present in the marsh, will change the food web of the ecosystem. Your investigation should determine which organisms the town should remove to limit the amount of change to the existing food web.

The guiding question of this investigation is, **Which member of an ecosystem would affect the food web the most if removed?**

Materials

You will use slides of marsh ecosystem organisms during your investigation. Slides are available from *www.nsta.org/adi-lifescience.*

Safety Precautions

Follow all normal lab safety rules.

Investigation Proposal Required?   ☐ Yes   ☐ No

Getting Started

Your teacher can provide you with a copy of slides that have information about the different organisms in the marsh ecosystem. Use these slides to analyze what changes might occur to the original food web for the marsh when any one of the organisms is removed from it.

To answer the guiding question, you must determine what type of data you need to collect, how you will collect it, and how you will analyze it. To determine *what type of data you need to collect*, think about the following questions:

- What information on the slides relates most to the food web of the marsh?

LAB 11

- How will you represent the data you use in different ways?
- What type of measurements or observations will you need to record during your investigation?

To determine *how you will analyze your data*, think about the following questions:

- How will you understand what the current food web looks like?
- Do you need to analyze all the different organisms, or should you focus on types of organisms?
- What type of graph could you create to help make sense of your data?

Connections to Crosscutting Concepts, the Nature of Science, and the Nature of Scientific Inquiry

As you work through your investigation, be sure to think about

- the use of models to study systems,
- how tracking the flow of energy and matter through systems allows scientists to understand these systems,
- how science is influenced by society, and
- the role of imagination and creativity when solving problems in science.

Initial Argument

Once your group has finished collecting and analyzing your data, you will need to develop an initial argument. Your argument must include a claim, evidence to support your claim, and a justification of the evidence. The claim is your group's answer to the guiding question. The evidence is an analysis and interpretation of your data. Finally, the justification of the evidence is why your group thinks the evidence matters. The justification of the evidence is important because scientists can use different kinds of evidence to support their claims. Your group will create your initial argument on a whiteboard. Your whiteboard should include all the information shown in Figure L11.2.

FIGURE L11.2

Argument presentation on a whiteboard

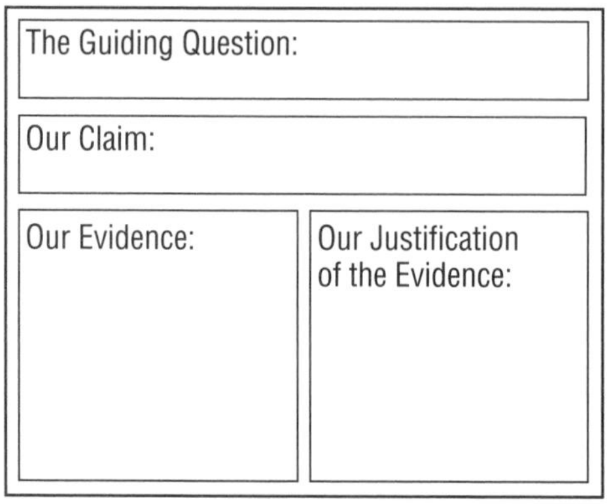

The Guiding Question:	
Our Claim:	
Our Evidence:	Our Justification of the Evidence:

Argumentation Session

The argumentation session allows all of the groups to share their arguments. One member of each group will stay at the lab station to share that group's argument, while the other members of the group go to the other lab stations one at a time to listen to and critique the arguments developed by their classmates. This is similar to how scientists present their arguments to other scientists at conferences. If you are responsible for critiquing your classmates' arguments, your goal is to look for mistakes so these mistakes can be fixed and they can make their argument better. The argumentation session is also a good time to think about ways you can make your initial argument better. Scientists must share and critique arguments like this to develop new ideas.

To critique an argument, you might need more information than what is included on the whiteboard. You will therefore need to ask the presenter lots of questions. Here are some good questions to ask:

- What did your group do to collect the data? Why do you think that way is the best way to do it?

- What did your group do to analyze the data? Why did your group decide to analyze it that way?

- What other ways of analyzing and interpreting the data did your group talk about?

- Why did your group decide to present your evidence in that way?

- What other claims did your group discuss before you decided on that one? Why did your group abandon those other ideas?

- How sure are you that your group's claim is accurate? What could you do to be more certain?

Once the argumentation session is complete, you will have a chance to meet with your group and revise your original argument. Your group might need to gather more data or design a way to test one or more alternative claims as part of this process. Remember, your goal at this stage of the investigation is to develop the most valid or acceptable answer to the research question!

Report

Once you have completed your research, you will need to prepare an investigation report that consists of three sections that provide answers to the following questions:

1. What question were you trying to answer and why?

2. What did you do during your investigation and why did you conduct your investigation in this way?

3. What is your argument?

Your report should answer these questions in two pages or less. The report must be typed, and any diagrams, figures, or tables should be embedded into the document. Be sure to write in a persuasive style; you are trying to convince others that your claim is acceptable or valid!

LAB 11

Lab 11. Food Webs and Ecosystems: Which Member of an Ecosystem Would Affect the Food Web the Most If Removed?

1. Imagine an ecosystem where mice eat the grass and foxes eat the mice. Explain what will happen to the population of foxes if there is a severe drought and all the grass dies.

2. The images below represent two food webs from different ecosystems that have similar animals. Which population of foxes would be least impacted by a drought that caused the grass to die in their ecosystem? Explain your reasoning.

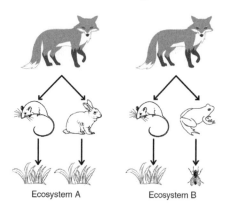

Ecosystem A   Ecosystem B

3. Society and culture often influence how scientists go about their work.

   a. I agree with this statement.

   b. I disagree with this statement.

Explain your answer, using an example from your investigation about food webs and ecosystems.

4. Scientists are very creative when they investigate the natural world.

    a. I agree with this statement.

    b. I disagree with this statement.

    Explain your answer, using an example from your investigation about food webs and ecosystems.

5. Scientists develop models to help them understand the natural world. Explain how a food web acts like a model and explain why such a model would be useful to scientists.

6. When scientists study the natural world, they often need to keep track of how matter and energy move through a system. Explain why understanding the flow of matter and energy is important, using an example from your investigation about food webs and ecosystems.

LAB 12

Lab Handout

Lab 12. Matter in Ecosystems: How Healthy Are Your Local Ecosystems?

Introduction

Ecosystems include all the living and nonliving things in a certain area. All living things in an ecosystem are called *biotic factors*. Biotic factors include the plants and animals in the ecosystem, as well as smaller organisms such as bacteria and fungi. All nonliving things in an ecosystem are referred to as *abiotic factors*. Abiotic factors include the water, soil, rocks, and air found in the ecosystem. Chemicals found in the ecosystem are also abiotic factors.

Water or air in an ecosystem is made up of many different chemicals. They include chemicals that are important for humans and animals to survive, such as the oxygen (O_2) we breathe and water molecules (H_2O). The air also contains other gases that other organisms use, such as carbon dioxide (CO_2) used by plants and nitrogen gas (N_2) used by bacteria. As living things use these chemicals for getting energy to survive, they will also release other chemicals that have similar elements, such as urea ($CO(NH_2)_2$) and phosphates (PO_4^{3-}). All of this activity means that different chemicals move between living and nonliving parts of an ecosystem.

The patterns of movement of matter through the living and nonliving parts of an ecosystem are known as *biogeochemical cycles*. The "bio" aspects of these cycles involve living things, and the "geo" aspects of these cycles involve nonliving things. All aspects of these cycles involve chemicals that are different forms of matter. The way scientists understand these cycles is based on a fundamental scientific law, the *law of conservation of matter*. This law tells us that as matter moves through an ecosystem, it will not be created or destroyed. The matter in an ecosystem simply changes forms as it is converted from one type of chemical to another.

Humans also add chemicals into different ecosystems. In many cases, these chemicals are used to help humans get what they need from the ecosystem. An example of this is fertilizer. Farmers use fertilizer containing nitrogen- and phosphorus-based chemicals to help crops grow. Chemicals that are added to an ecosystem can benefit some organisms while also harming others.

Keeping natural cycles in balance helps keep ecosystems healthy. One way scientists figure out how healthy ecosystems are is through testing living and nonliving parts of them. It is a little more difficult to test living organisms in an ecosystem than it is to test nonliving parts of an ecosystem. Scientists will often test soil and water and other nonliving parts of an ecosystem, to understand how different chemicals are moving through an ecosystem.

Using this information, scientists can determine if there is too much or too little of a certain chemical in an ecosystem. Looking at the levels of different chemicals helps them figure out how healthy the ecosystem is. The amounts of similar chemicals in different locations (water or soil) can show how matter is moving through an ecosystem.

Your Task

Conduct an investigation of ecosystems where you live and how healthy they are. Look at ecosystems that include both soil and water sources.

The guiding question of this investigation is, **How healthy are your local ecosystems?**

Materials

You may use any of the following materials during your investigation:

Consumables	Equipment
• Samples of soil from three locations in your area • Samples of water from three locations in your area	• Soil quality test kit (nitrogen, oxygenation, phosphorus, pH) • Water quality test kit (pH, nitrates, phosphates, dissolved O_2, turbidity) • Sanitized indirectly vented chemical-splash goggles • Chemical-resistant apron • Gloves • Lab 12 Reference Sheet

Safety Precautions

Follow all normal lab safety rules. In addition, take the following safety precautions:

1. Put on sanitized indirectly vented chemical-splash goggles and laboratory apron and gloves before starting the lab activity.

2. Handle all glassware with care to avoid breakage. Sharp glass edges can cut skin!

3. Review the important information on chemicals on the safety data sheet, and use caution when handling chemicals.

4. Wash hands with soap and water after completing the lab activity.

Investigation Proposal Required?   ☐ Yes   ☐ No

Getting Started

To answer the guiding question, you will need to analyze water and soil samples taken from local ecosystems and use background information about chemical cycling in ecosystems. To accomplish this task, you must first determine what type of data you need to collect, how you will collect it, and how you will analyze it. To determine *what type of data you need to collect*, think about the following questions:

- What type of information do I need to collect from the Lab 12 Reference Sheet?
- What type of tests do I need to determine the quality of the water and soil samples? (*Hint:* Be sure to follow all directions given in the water and soil test kits.)
- What type of measurements or observations will you need to record during your investigation?

To determine *how you will collect your data*, think about the following questions:

- What will serve as a control (or comparison) condition?
- How will you make sure that your data are of high quality (i.e., how will you reduce error)?
- How will you keep track of the data you collect and how will you organize the data?

To determine *how you will analyze your data*, think about the following questions:

- What type of calculations will you need to make?
- What type of graph could you create to help make sense of your data?

Connections to Crosscutting Concepts, the Nature of Science, and the Nature of Scientific Inquiry

As you work through your investigation, be sure to think about

- how energy and matter flow into, out of, within, and through a system;
- how changes to different parts of ecosystems affect their stability;
- how observations and inferences are different but related to each other; and
- the difference between data collected in an investigation and evidence created in an investigation.

Initial Argument

Once your group has finished collecting and analyzing your data, you will need to develop an initial argument. Your argument must include a claim, evidence to support your claim, and a justification of the evidence. The claim is your group's answer to the guiding question. The evidence is an analysis and interpretation of your data. Finally, the justification of the evidence is why your group thinks the evidence matters. The justification of the evidence is important because scientists can use different kinds of evidence to support their claims. Your group will create your initial argument on a whiteboard. Your whiteboard should include all the information shown in Figure L12.1.

Argumentation Session

The argumentation session allows all of the groups to share their arguments. One member of each group will stay at the lab station to share that group's argument, while the other members of the group go to the other lab stations one at a time to listen to and critique the arguments developed by their classmates. This is similar to how scientists present their arguments to other scientists at conferences. If you are responsible for critiquing your classmates' arguments, your goal is to look for mistakes so these mistakes can be fixed and they can make their argument better. The argumentation session is also a good time to think about ways you can make your initial argument better. Scientists must share and critique arguments like this to develop new ideas.

FIGURE L12.1 _____

Argument presentation on a whiteboard

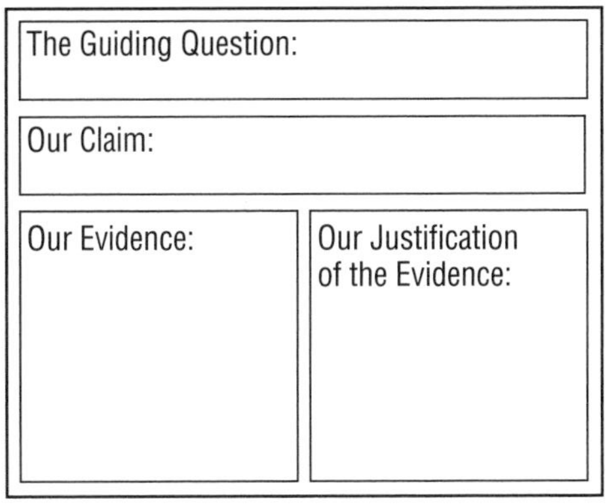

The Guiding Question:	
Our Claim:	
Our Evidence:	Our Justification of the Evidence:

To critique an argument, you might need more information than what is included on the whiteboard. You will therefore need to ask the presenter lots of questions. Here are some good questions to ask:

- What did your group do to collect the data? Why do you think that way is the best way to do it?
- What did your group do to analyze the data? Why did your group decide to analyze it that way?
- What other ways of analyzing and interpreting the data did your group talk about?
- What did your group do to make sure that these calculations are correct?
- Why did your group decide to present your evidence in that way?
- What other claims did your group discuss before you decided on that one? Why did your group abandon those other ideas?
- How sure are you that your group's claim is accurate? What could you do to be more certain?

Once the argumentation session is complete, you will have a chance to meet with your group and revise your original argument. Your group might need to gather more data or design a way to test one or more alternative claims as part of this process. Remember, your goal at this stage of the investigation is to develop the most valid or acceptable answer to the research question!

Report

Once you have completed your research, you will need to prepare an investigation report that consists of three sections that provide answers to the following questions:

LAB 12

1. What question were you trying to answer and why?

2. What did you do during your investigation and why did you conduct your investigation in this way?

3. What is your argument?

Your report should answer these questions in two pages or less. The report must be typed, and any diagrams, figures, or tables should be embedded into the document. Be sure to write in a persuasive style; you are trying to convince others that your claim is acceptable or valid!

Lab 12 Reference Sheet

The Nitrogen Cycle and the Phosphorus Cycle

Figure R12.1 shows the different forms that nitrogen gets combined into as it moves through an ecosystem. Nitrogen is a basic chemical needed for life. One of the most important functions of nitrogen is being one of the elements that form DNA. DNA is the molecule that allows living things to grow and reproduce. Nitrogen is also an important piece for building proteins. All living things in an ecosystem have nitrogen, and nitrogen moves from living to nonliving parts of an ecosystem in different forms and ways.

FIGURE R12.1

The nitrogen cycle

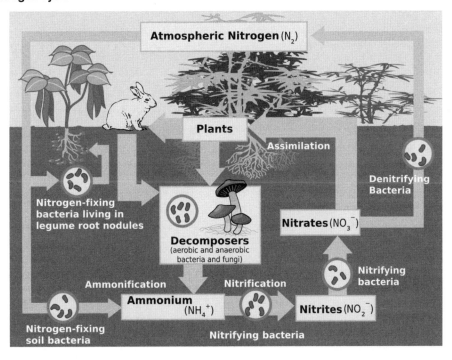

Plants and animals move nitrogen into air, water, and soil through the waste products they make. They also release nitrogen into an ecosystem when they die and their bodies decompose. Decomposing involves the breakdown of organic material into smaller chemicals. Different kinds of bacteria are very important in changing the forms of nitrogen that move through the nonliving parts of an ecosystem. Let's look at the different forms of nitrogen found in the nitrogen cycle.

Nitrogen gas (N_2): Nitrogen gas is the most common chemical in the air we breathe. However, plants and animals do not usually use nitrogen gas. Certain bacteria in soil can convert nitrogen gas into forms that plants and other bacteria can use. These bacteria are called *nitrogen-fixing*.

Ammonium (NH_4^+): Ammonium is a chemical you may have heard of in household cleaners. Ammonium moves into the soil when nitrogen-fixing bacteria change N_2 into NH_4^+. Ammonium is also produced when living things decompose. Ammonium in soil can be absorbed by plants through their roots and by other kinds of bacteria. Plants use ammonium as a source for nitrogen in their proteins and DNA. High levels of ammonium in water can poison fish and other living things there.

Nitrite (NO_2^-): Nitrite is a very unstable form for nitrogen in soil. Certain bacteria, called *nitrifying bacteria*, convert ammonium in the soil into nitrite. However, nitrite is usually changed into another form known as nitrate (see the next paragraph). This change is also done by nitrifying bacteria. Nitrite can be found in water sources. High levels of nitrite can cause diseases in some fish species.

Nitrate (NO_3^-): Nitrate is a more stable nitrogen molecule found in both soil and water. Nitrates can be absorbed by plants through their roots. Plants also use nitrates as a source for nitrogen in their proteins and DNA. Nitrates can be converted back into N_2 through another kind of bacteria called *denitrifying bacteria*. Nitrates are typically found in both soil and water. Nitrate levels in water are important to monitor. High levels of nitrates in drinking water can lead to serious health conditions in babies. High nitrate levels in water can also help algae grow out of control. Overgrowing algae can decrease the amount of nutrients, like oxygen, that other plants and animals use. As the algae continue to grow, the other plants and animals will die.

Figure R12.2 shows how phosphorus moves through an ecosystem, usually in the form of phosphate. Phosphate is an important molecule that living things use to build their DNA. Phosphorus is not typically found in the air in an ecosystem. Phosphate is mostly found in rocks and soil. Phosphate present in soil can be absorbed by plants to use. Animals eat the plants to get the phosphate they need. Rocks can be worn away by water rushing over them for long periods of time. As the water wears away the rock, phosphate is released into rivers, streams, and lakes. Animals and plants in the water can absorb the phosphate for their use. Phosphate that is not used by living things will fall to the bottom of the bodies of water as sediment. Over long periods of time, the phosphate sediment will change form into rocks again. Many fertilizers used by farmers contain high levels of phosphate. Phosphate helps algae grow out of control in fresh water. Overgrowing algae can decrease the amount of nutrients, like oxygen, that other plants and animals use. As the algae continue to grow, the other plants and animals will die.

FIGURE R12.2

The phosphorus cycle

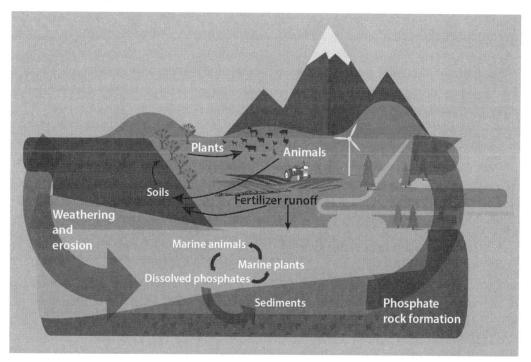

Checkout Questions

Lab 12. Matter in Ecosystems: How Healthy Are Your Local Ecosystems?

1. Farmer Johnson has two small ponds on his property. The first pond is next to a cornfield near the top of a hill. It has lots of fish and some vegetation. The second pond is also near the cornfield, but farther down the hill. The second pond has no fish and is almost full of vegetation. Use what you know about how matter moves through an ecosystem to explain why the two ponds have very different characteristics.

2. Organic farming involves growing fruits and vegetables without using any chemicals on the crops like pesticides (chemicals that kill bugs) or herbicides (chemicals that kill unwanted weeds). Use what you know about matter and how it moves through an ecosystem to explain why organic farming is a popular choice.

3. Observations are more important in science than inferences.

    a. I agree with this statement.

    b. I disagree with this statement.

    Explain your answer, using an example from your investigation about matter in ecosystems.

4. In science, data and evidence are the same thing.

   a. I agree with this statement.

   b. I disagree with this statement.

   Explain your answer, using an example from your investigation about matter in ecosystems.

5. When scientists study the natural world, they often need to keep track of how matter and energy moves through a system. Explain why understanding the flow of matter and energy is important, using an example from your investigation about matter in ecosystems.

6. Scientists study complex systems that have many related parts. Change in one aspect of a system can have impacts on many other parts of the system. Explain how changes in one part of an ecosystem can influence other parts of the system, using an example from your investigation about matter in ecosystems.

Lab Handout

Lab 13. Carbon Cycling: Which Carbon Cycle Process Affects Atmospheric Carbon the Most?

Introduction

Organisms live together in ecosystems and rely on each other for food. All living things also require different amounts and kinds of *nutrients*, including nonliving factors in an ecosystem. Nutrients are chemicals that are essential for plant and animal growth. Animals get nutrients by eating plants or other animals. Plants get nutrients from the nonliving parts of an ecosystem, such as water and soil. In fact, nutrients often cycle through an ecosystem, which means they change forms and move through both living and nonliving parts. Carbon, nitrogen, and phosphorus are chemicals that cycle through an ecosystem. The carbon cycle is one of the most important cycles that support life in an ecosystem. Figure L13.1 illustrates the sources and storage of carbon as it cycles through an ecosystem.

FIGURE L13.1 _____

The carbon cycle in ecosystems

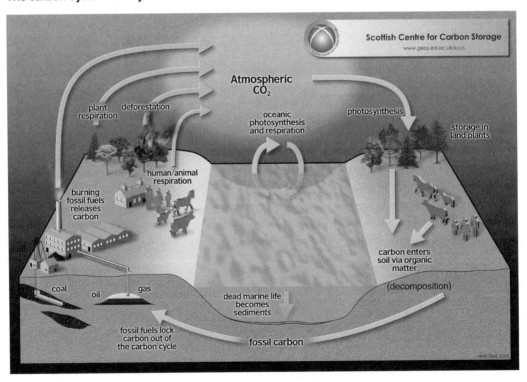

The following list traces the path of a carbon atom:

- Atmospheric carbon dioxide gas (CO_2) is a source of carbon in the cycle. The atmosphere is the layer of gases that surround the planet. CO_2 passes into the ecosystem through living organisms and nonliving elements through different processes.

- CO_2 leaves the atmosphere through *photosynthesis*, which is the chemical process used by plants to make food (in the form of sugar) using CO_2 from the atmosphere. This carbon material is eventually moved into other organisms through the food web. Forests and other areas with a lot of plant life are where the greatest amounts of photosynthesis happen.

- CO_2 can be released into the atmosphere through *combustion*. When trees are burned, CO_2 is released. The burning of large areas of trees is known as *deforestation*. Human activities add more CO_2 into the atmosphere through activities like the burning of *fossil fuels*, which are carbon-rich materials made from the broken-down remains of plants and animals that lived millions of years ago. These fuels include coal, oil, and natural gas. Humans use these fuels to provide energy for technology and communities.

There can be more CO_2 in the atmosphere than plants can use in photosynthesis. This extra carbon in the atmosphere can lead to several changes in the environment. Extra CO_2 can trap heat energy coming from the planet's surface. As this heat energy gets trapped, it can raise the global temperature. This heating due to extra CO_2 is known as the *greenhouse effect*. Scientists have demonstrated that the rise in global temperature leads to changes in the climate of different regions of the world. *Climate* includes trends in different conditions in the environment that happen over long periods of time. Extra atmospheric CO_2 can also dissolve into bodies of water, especially oceans. When CO_2 combines with water, carbonic acid (H_2CO_3) is formed. Over time, this process leads to lower pH levels in the ocean, which is called *ocean acidification*.

Your Task

Use a computer simulation to determine whether atmospheric CO_2 will decrease more from burning fossil fuels or from changing the amount of deforestation.

The guiding question of this investigation is, **Which carbon cycle process affects atmospheric carbon the most?**

Materials

You will use an online simulation called *Carbon Lab* to conduct your investigation. You can find the simulation by going to the following website: *www.learner.org/courses/envsci/interactives/carbon/carbon.html*.

LAB 13

Safety Precautions

Follow all normal lab safety rules.

Investigation Proposal Required?   ☐ Yes   ☐ No

Getting Started

Use the *Carbon Lab* simulation to estimate how decreasing plant life through deforestation and increasing the fossil fuel use will affect other parts of the planet. You will need to run simulations using different settings for increasing and decreasing the amount of fossil fuels burned. You will also need to run simulations using different settings for increasing and decreasing the amount of deforestation happening. You can adjust these settings using the sliding scales on the screen. Once you have set the scales, then you need to press the "Run Decade" button. One decade will show what happens to atmospheric CO_2 after 10 years at the settings you choose. The image will also show how CO_2 is distributed in different parts of the environment after that decade as well. You will need to run many decades to see what happens over longer periods of time. You may notice that the units of CO_2 provided are GT, which stands for gigatons; 1 GT is equal to 1,000,000,000,000,000 grams of a substance.

To answer the guiding question, you will need to observe changes in the amount of atmospheric carbon. You can also observe how much carbon is stored in different parts of the environment. To accomplish this task, you must first determine what type of data you need to collect, how you will collect it, and how you will analyze it. To determine *what type of data you will need to collect*, think about the following questions:

- What type of data is available in the simulation?
- What information about carbon cycle processes is provided by different data sources?
- What type of measurements or observations will you need to record during your investigation?

To determine *how you will collect your data*, think about the following questions:

- What will serve as a control (or comparison) condition?
- How will you make sure that your data are of high quality (i.e., how will you reduce error)?
- How will you keep track of the data you collect and how will you organize the data?

To determine *how you will analyze your data*, think about the following questions:

- What type of calculations will you need to make?
- What type of graph could you create to help make sense of your data?

Connections to Crosscutting Concepts, the Nature of Science, and the Nature of Scientific Inquiry

As you work through your investigation, be sure to think about

- how events of scientific interest occur over different scales of time,
- how changes to different parts of ecosystems affect their stability,
- the difference between data collected in an investigation and evidence created in an investigation, and
- how scientists use different methods in their investigations.

Initial Argument

Once your group has finished collecting and analyzing your data, you will need to develop an initial argument. Your argument must include a claim, evidence to support your claim, and a justification of the evidence. The claim is your group's answer to the guiding question. The evidence is an analysis and interpretation of your data. Finally, the justification of the evidence is why your group thinks the evidence matters. The justification of the evidence is important because scientists can use different kinds of evidence to support their claims. Your group will create your initial argument on a whiteboard. Your whiteboard should include all the information shown in Figure L13.2.

FIGURE L13.2 _____

Argument presentation on a whiteboard

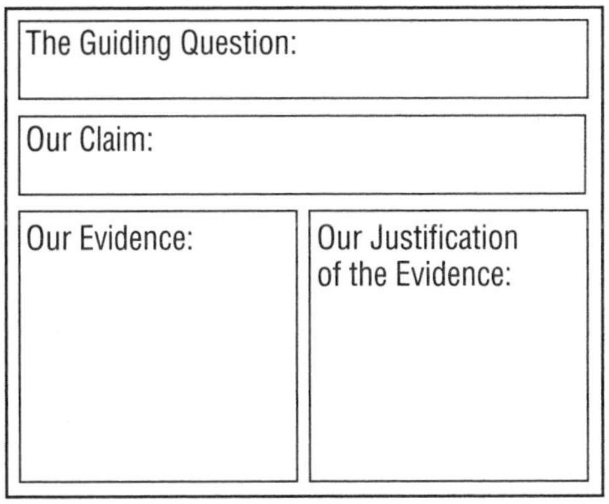

The Guiding Question:	
Our Claim:	
Our Evidence:	Our Justification of the Evidence:

Argumentation Session

The argumentation session allows all of the groups to share their arguments. One member of each group will stay at the lab station to share that group's argument, while the other members of the group go to the other lab stations one at a time to listen to and critique the arguments developed by their classmates. This is similar to how scientists present their arguments to other scientists at conferences. If you are responsible for critiquing your classmates' arguments, your goal is to look for mistakes so these mistakes can be fixed and they can make their argument better. The argumentation session is also a good time to think about ways you can make your initial argument better. Scientists must share and critique arguments like this to develop new ideas.

To critique an argument, you might need more information than what is included on the whiteboard. You will therefore need to ask the presenter lots of questions. Here are some good questions to ask:

- What did your group do to collect the data? Why do you think that way is the best way to do it?

- What did your group do to analyze the data? Why did your group decide to analyze it that way?
- What other ways of analyzing and interpreting the data did your group talk about?
- What did your group do to make sure that these calculations are correct?
- Why did your group decide to present your evidence in that way?
- What other claims did your group discuss before you decided on that one? Why did your group abandon those other ideas?
- How sure are you that your group's claim is accurate? What could you do to be more certain?

Once the argumentation session is complete, you will have a chance to meet with your group and revise your original argument. Your group might need to gather more data or design a way to test one or more alternative claims as part of this process. Remember, your goal at this stage of the investigation is to develop the most valid or acceptable answer to the research question!

Report

Once you have completed your research, you will need to prepare an investigation report that consists of three sections that provide answers to the following questions:

1. What question were you trying to answer and why?

2. What did you do during your investigation and why did you conduct your investigation in this way?

3. What is your argument?

Your report should answer these questions in two pages or less. The report must be typed, and any diagrams, figures, or tables should be embedded into the document. Be sure to write in a persuasive style; you are trying to convince others that your claim is acceptable or valid!

Checkout Questions

Lab 13. Carbon Cycling: Which Carbon Cycle Process Affects Atmospheric Carbon the Most?

1. Zachary thinks that deforestation only affects the levels of carbon dioxide gas (CO_2) in the atmosphere because of burning the trees that get cut down. Elizabeth thinks cutting the trees down is enough to affect the levels of CO_2 in the atmosphere. Using what you know about the carbon cycle, provide supporting evidence for either Zachary or Elizabeth.

2. Some students in science class were discussing photosynthesis and the carbon cycle. The students were confused about the impact of burning fossil fuels. Burning fossil fuels releases CO_2 into the atmosphere, and plants use CO_2 during photosynthesis to grow. The students concluded that burning fossil fuels is a good thing because all the plants and trees will grow bigger. Use what you know about the carbon cycle and atmospheric CO_2 to help clear up their confusion.

3. In science, data and evidence are the same thing.

    a. I agree with this statement.

    b. I disagree with this statement.

Explain your answer, using an example from your investigation about carbon cycling.

4. When scientists study complex systems like the carbon cycle, they must focus on processes that occur on very different time scales. For example, trees can take 100 years to reach full size, fossil fuels have formed over millions of years, photosynthesis is an ongoing process, and trees can be cut down and burned in just a few seconds or minutes. Explain why understanding the different time scales for events is important, using an example from your investigation about carbon cycling.

5. Scientists study complex systems that have many related parts. Change in one aspect of a system can have impacts on many other parts of the system. Explain how changes in one aspect of the carbon cycle influence other aspects of the system.

SECTION 4
Life Sciences Core Idea 3

Heredity: Inheritance and Variation in Traits

Introduction Labs

LAB 14

Lab Handout

Lab 14. Variation in Traits: How Do Beetle Traits Vary Within and Across Species?

Introduction

Organisms differ from one another in several ways. When those differences are so great that organisms are unable to mate and produce fertile offspring, they are said to be of different *species*. Differences between species, or so-called *interspecies differences*, can be great or small. Some differences can be easily seen, such as different shapes of arms or legs. Others are not easy to observe, such as differences in the kinds of genes each species has. Differences also exist within a species; in other words, not all members of the same species are the same—for example, not all gray whales are the same. Indeed, just look around you. All of the students in this class look very different even though they are all members of the same species. Some of these differences are unique to individual organisms or a group of organisms, especially in some physical traits, like special markings. Other differences, like ones in types of genes or behaviors, are more common among a particular species but make them distinct from other species. Some differences among species help certain forms survive better in certain environments than others.

These differences and similarities, both within and across species, are known as *variation*. To understand the variation present in living things, scientists developed a system of classification—that is, grouping organisms together based on their differences and similarities. This system begins with a few large groups of organisms that share a few similar traits on the first level. The next level of classification splits those large groups into smaller groups using differences and similarities among the organisms in them. This second set of groups is then split up even further based on the variation among the organisms. Current classification systems use eight major levels to classify all the living things. Some of these levels also have sublevels, which further improves classification. *Taxonomy* is the branch of science that deals with classifying organisms.

Take the beetle as an example of this classification system. The name Beetle is given to an order of insects that all have elytra. The elytra are a hardened, sheathlike set of front wings, which usually cover the entire abdomen when the insect is not in flight. Beetles vary greatly in size. The largest is the Eastern Hercules beetle (*Dynastes tityus*), which grows to 16 cm (up to 6.3 inches) in length (Figure L14.1). Other species may be less than 0.1 cm (less than 0.04 inch) long.

The Beetle order is split into a number of smaller groups, which involve more similarities in shape between the group members. These smaller groups follow a strict hierarchy. The major levels and sublevels smaller than order are called suborder, family, subfamily,

FIGURE L14.1 _____

Eastern Hercules beetle

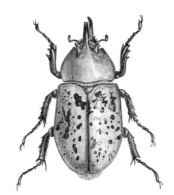

FIGURE L14.2 _____

European violet ground beetle

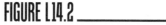

and genus. A genus is the smallest group of importance in the naming of individual species, although in some classifications generic groups may be further split into subgenera. The scientific name of a species includes, first, its genus and, second, its specific name. For example, the European violet ground beetle (Figure L14.2) is called *Carabus violaceus*, meaning the species *violaceus* in the genus *Carabus*. (Genus and species names are always printed in italics. The genus name is capitalized, and the species name is lowercase [starts with a small letter].) The full classification of this insect is shown in Table L14.1.

TABLE L14.1 _____

Classification of the European violet ground beetle

Classification level	Name	Description
Phylum	Arthropoda	Arthropod
Class	Insecta	Insect
Subclass	Pterygota	Winged insect
Order	Coleoptera	Beetle
Suborder	Adephaga	Carnivorous beetle
Family	Carabidae	Ground beetle
Genus	*Carabus*	*
Species	*violaceus*	Violet ground beetle

The Beetle order embraces more species than any other group in the animal kingdom. At least 250,000 species are known. This represents more than one-quarter of all animal species and is far more than all mammal species combined. With all these different species, a lot of variation exists among beetles.

LAB 14

Your Task

Examine the amount and nature of the variation that exists within and across the three different species of beetle shown in the Lab Reference Sheet: The three species you can examine in the information sheets are the ground beetle (*Harpalus affinis*), the figeater beetle (*Cotinis mutabilis*), and the potato beetle (*Leptinotarsa decemlineata*).

The guiding question of this investigation is, **How do beetle traits vary within and across species?**

Materials

You may use any of the following materials during your investigation:

- Lab 14 Reference Sheet with beetle information, or preserved samples of beetles
- Insect information websites such as *www.bugguide.net* and *http://species.wikimedia.org*

Safety Precautions

Follow all normal lab safety rules.

Investigation Proposal Required?   ☐ Yes   ☐ No

Getting Started

To answer the guiding question, you will need to observe variations between different examples of beetles within each species and across the species provided. To accomplish this task, you must first determine what type of data you need to collect, how you will collect it, and how you will analyze it. To determine *what type of data you will need to collect*, think about the following questions:

- What type of data can you collect from the Lab Reference Sheet?
- What information about the beetles are available from internet sources?
- What type of measurements or observations will you need to record during your investigation?

To determine *how you will collect your data*, think about the following questions:

- What will serve as a control (or comparison) condition?
- How will you make sure that your data are of high quality (i.e., how will you reduce error)?
- How will you keep track of the data you collect and how will you organize the data?

To determine *how you will analyze your data*, think about the following questions:

- What type of calculations will you need to make?
- What type of graph could you create to help make sense of your data?

Connections to Crosscutting Concepts, the Nature of Science, the Nature of Scientific Inquiry

As you work through your investigation, be sure to think about

- the importance of looking for patterns in science,
- how organisms' structures are related to the functions they perform,
- the different roles observations and inferences play in science, and
- how scientific knowledge changes over time.

Initial Argument

Once your group has finished collecting and analyzing your data, you will need to develop an initial argument. Your argument must include a claim, evidence to support your claim, and a justification of the evidence. The claim is your group's answer to the guiding question. The evidence is an analysis and interpretation of your data. Finally, the justification of the evidence is why your group thinks the evidence matters. The justification of the evidence is important because scientists can use different kinds of evidence to support their claims. Your group will create your initial argument on a whiteboard. Your whiteboard should include all the information shown in Figure L14.3.

FIGURE L14.3

Argument presentation on a whiteboard

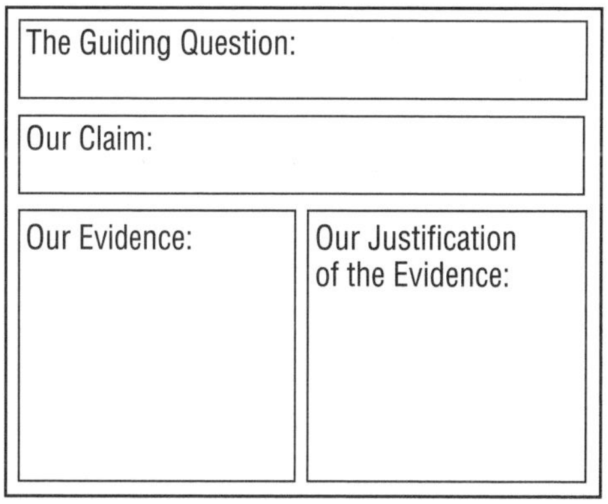

The Guiding Question:	
Our Claim:	
Our Evidence:	Our Justification of the Evidence:

Argumentation Session

The argumentation session allows all of the groups to share their arguments. One member of each group will stay at the lab station to share that group's argument, while the other members of the group go to the other lab stations one at a time to listen to and critique the arguments developed by their classmates. This is similar to how scientists present their arguments to other scientists at conferences. If you are responsible for critiquing your classmates' arguments, your goal is to look for mistakes so these mistakes can be fixed and they can make their argument better. The argumentation session is also a good time to think about ways you can make your initial argument better. Scientists must share and critique arguments like this to develop new ideas.

To critique an argument, you might need more information than what is included on the whiteboard. You will therefore need to ask the presenter lots of questions. Here are some good questions to ask:

- What did your group do to collect the data? Why do you think that way is the best way to do it?
- What did your group do to analyze the data? Why did your group decide to analyze it that way?
- What other ways of analyzing and interpreting the data did your group talk about?
- Why did your group decide to present your evidence in that way?
- What other claims did your group discuss before you decided on that one? Why did your group abandon those other ideas?
- How sure are you that your group's claim is accurate? What could you do to be more certain?

Once the argumentation session is complete, you will have a chance to meet with your group and revise your original argument. Your group might need to gather more data or design a way to test one or more alternative claims as part of this process. Remember, your goal at this stage of the investigation is to develop the most valid or acceptable answer to the research question!

Report

Once you have completed your research, you will need to prepare an investigation report that consists of three sections that provide answers to the following questions:

1. What question were you trying to answer and why?

2. What did you do during your investigation and why did you conduct your investigation in this way?

3. What is your argument?

Your report should answer these questions in two pages or less. The report must be typed, and any diagrams, figures, or tables should be embedded into the document. Be sure to write in a persuasive style; you are trying to convince others that your claim is acceptable or valid!

Lab 14 Reference Sheet
Three Types of Beetles

Ground beetle (*Harpalus affinis*)

a.

b.

c.

- Locations: Europe, Asia, North America, Middle East, and Australia
- Size: 8.5–12 mm
- Habitats: dry areas such as open farmland, parks, gardens, and sand dunes
- Reproduction
  - Spring is the main egg-laying season, but some egg laying occurs in summer.
  - Larvae and adults are present from winter into spring.

Figeater beetle (*Cotinis mutabilis*)

a.

b.

c.

- Locations: southwest United States, Mexico
- Size: 3 cm (adult)
- Habitats: areas where they can feed on pollen and nectar, as well as damaged fruits
- Reproduction:
  - Spring is the main time for transition from larvae to adults.
  - Adults emerge July–September.
  - Eggs and larvae stage occur over winter.

Potato beetle (*Leptinotarsa decemlineata*)

a.

b.

c.

- Locations: southwest United States, Mexico, Europe, and southern Russia
- Size: 10 mm (adult)
- Habitats: found mainly near farmland (crop pest for potato agriculture)
- Reproduction
    - Very quick, typically one month to go from egg to adult
    - Eggs laid on the underside of leaves
    - Temperature and light dependent
    - Three generations can grow during a single crop season

LAB 14

Lab 14. Variation in Traits: How Do Beetle Traits Vary Within and Across Species?

1. Describe how variation exists among different species that are similar.

2. Physical and genetic differences exist among the many types of living organisms that exist on the planet. Explain how scientists use the variation they observe among different species to classify them in different size groups. Does more variation exist among organisms that are in the same class or the same family? Explain your answer.

3. Observations are the same thing in science as inferences.

   a. I agree with this statement.
   b. I disagree with this statement.

   Explain your answer, using an example from your investigation about variation in traits.

4. Scientific knowledge can change over time but is still reliable.

   a. I agree with this statement.
   b. I disagree with this statement.

   Explain your answer, using an example from your investigation about variation in traits.

5. Scientists often look for patterns when they study nature. Explain why understanding patterns in nature are important, using an example from your investigation about variation in traits.

6. The relationship between the structure and function of organisms' features is an important area of study in science. Discuss why it is important for scientists to understand this relationship, using an example from your investigation about variations in traits.

Lab Handout

Lab 15. Mutations in Genes: How Do Different Types of Mutations in Genes Affect the Function of an Organism?

FIGURE L15.1

Section of a DNA molecule

FIGURE L15.2

Connections between DNA, RNA, and proteins

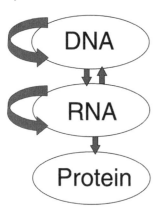

Introduction

During reproduction, information for traits is passed from the parent organisms to the offspring. This transfer of trait information from parents to offspring is known as *inheritance*. The factors with the trait information are found in all living things and are called *genes*. All of the genes passed on during reproduction contain information needed to create a new organism. Genes are made of a molecule known as *DNA*, which stands for deoxyribonucleic acid. Figure L15.1 shows a section of a DNA molecule. DNA is made of two strands of molecules with a sugar/phosphate side and a base side. The base sides interact with the base sides of another strand to connect the two strands together. There are only four types of bases in DNA, called *A, G, C,* and *T*. These four bases will bond in only two ways: an A on one strand will only bond to a T on the other strand, and a C on one strand will only bond to a G on the other strand. As these bases pair up, the two strands stay connected, forming larger molecules of DNA.

The chemical structure of the DNA molecule allows it to store information. Genes are made up of long units of DNA. The order of the bases in a strand contains a code for the structure of other molecules in organisms. A section of a strand of DNA, like TACCGATGATTCCGG, has a code that tells an organism's cells how to build other molecules. The DNA in genes is read by special molecules, called enzymes, which use the code in DNA to build an *RNA* molecule. RNA, which stands for ribonucleic acid, is a single-stranded molecule similar to DNA that also contains a sequence of bases. The RNA molecule made from a specific gene is then used to make a *protein* molecule (see the next paragraph). Figure L15.2 shows the connections between DNA, RNA, and proteins.

Proteins are molecules that perform all kinds of functions in cells. A protein's function is determined by the way it is shaped, and that shape is determined by the order of *amino acids* that make up a protein's structure. Amino acids are small molecules that make up larger protein molecules, much like the sugar/phosphate + base molecules make up DNA and RNA. Thus, in all of these large molecules, the order structure of the smaller molecules that form them determine the functions the larger molecules perform.

Changes to the order of those small molecules are called *mutations*. There are several types of mutations that can occur to the order of bases in DNA. *Substitution mutations* happen when pairs of bonded bases in a double strand of DNA get replaced with different pairs. An example of a substitution mutation is having an A-T base pair replaced by a G-C pair or a T-A pair. *Insertion mutations* happen when extra base pairs are included in an existing DNA sequence. *Deletion mutations* happen when base pairs in an existing DNA molecule get removed from it. Figure L15.3 shows each type of mutation using the same original DNA sequence. For all of these mutations, the location in the order of bases in DNA is important. Some mutations will not change the amino acid coded for by a specific sequence, but others will.

FIGURE L15.3

Types of DNA-level mutations

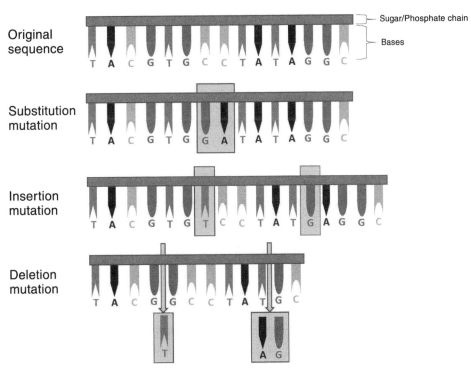

LAB 15

Your Task

Examine the effects of different types of mutations in a DNA sequence on the resulting RNA and protein molecules. Since these molecules are not easy to work with in a classroom, you will be using a computer simulation to investigate the effect of mutations.

The guiding question of this investigation is, **How do different types of mutations in genes affect the function of an organism?**

ials

ill use an online simulation called *Mutations* to conduct your investigation. You can find nulation by going to the following website: *http://concord.org/stem-resources/mutations.*

Safety Precautions

Follow all normal lab safety rules.

Investigation Proposal Required?   ☐ Yes   ☐ No

Getting Started

To answer the guiding question, you will use a simulation to observe the impact of different mutations on the resulting protein. To accomplish this task, you must first determine what type of data you need to collect, how you will collect it, and how you will analyze it. To determine *what type of data you will need to collect*, think about the following questions:

- What type of data can you collect from the simulation?
- What types of mutations can you make using the simulation?
- What type of measurements or observations will you need to record during your investigation?

To determine *how you will collect your data*, think about the following questions:

- What will serve as a control (or comparison) condition?
- How will you make sure that your data are of high quality (i.e., how will you reduce error)?
- How will you keep track of the data you collect and how will you organize the data?

To determine *how you will analyze your data*, think about the following questions:

- What type of calculations will you need to make?
- What type of graph could you create to help make sense of your data?

National Science Teachers Association

Connections to Crosscutting Concepts, the Nature of Science, and the Nature of Scientific Inquiry

As you work through your investigation, be sure to think about

- how different actions in living things happen on different scales of size and time,
- how organisms' structures are related to the functions they perform,
- the different roles theories and laws play in science, and
- how imagination and creativity are necessary for developing scientific knowledge.

Initial Argument

Once your group has finished collecting and analyzing your data, you will need to develop an initial argument. Your argument must include a claim, evidence to support your claim, and a justification of the evidence. The claim is your group's answer to the guiding question. The evidence is an analysis and interpretation of your data. Finally, the justification of the evidence is why your group thinks the evidence matters. The justification of the evidence is important because scientists can use different kinds of evidence to support their claims. Your group will create your initial argument on a whiteboard. Your whiteboard should include all the information shown in Figure L15.4.

FIGURE L15.4
Argument presentation on a whiteboard

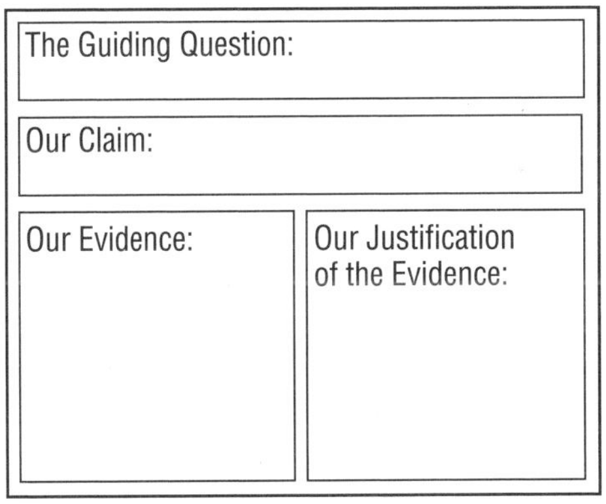

Argumentation Session

The argumentation session allows all of the groups to share their arguments. One member of each group will stay at the lab station to share that group's argument, while the other members of the group go to the other lab stations one at a time to listen to and critique the arguments developed by their classmates. This is similar to how scientists present their arguments to other scientists at conferences. If you are responsible for critiquing your classmates' arguments, your goal is to look for mistakes so these mistakes can be fixed and they can make their argument better. The argumentation session is also a good time to think about ways you can make your initial argument better. Scientists must share and critique arguments like this to develop new ideas.

To critique an argument, you might need more information than what is included on the whiteboard. You will therefore need to ask the presenter lots of questions. Here are some good questions to ask:

- What did your group do to collect the data? Why do you think that way is the best way to do it?

- What did your group do to analyze the data? Why did your group decide to analyze it that way?

- What other ways of analyzing and interpreting the data did your group talk about?

- Why did your group decide to present your evidence in that way?

- What other claims did your group discuss before you decided on that one? Why did your group abandon those other ideas?

- How sure are you that your group's claim is accurate? What could you do to be more certain?

Once the argumentation session is complete, you will have a chance to meet with your group and revise your original argument. Your group might need to gather more data or design a way to test one or more alternative claims as part of this process. Remember, your goal at this stage of the investigation is to develop the most valid or acceptable answer to the research question!

Report

Once you have completed your research, you will need to prepare an investigation report that consists of three sections that provide answers to the following questions:

1. What question were you trying to answer and why?

2. What did you do during your investigation and why did you conduct your investigation in this way?

3. What is your argument?

Your report should answer these questions in two pages or less. The report must be typed, and any diagrams, figures, or tables should be embedded into the document. Be sure to write in a persuasive style; you are trying to convince others that your claim is acceptable or valid!

Checkout Questions

Lab 15. Mutations in Genes: How Do Different Types of Mutations in Genes Affect the Function of an Organism?

1. Describe the different kinds of mutations that can happen in a sequence of DNA.

2. A muscle cell in an organism's body has stopped making one of the proteins needed to make a muscle fiber. That cell was exposed to a large amount of radiation previously. Explain how mutations could have led to this problem in this muscle cell.

3. Theories and laws serve the same purpose in science.

   a. I agree with this statement.
   b. I disagree with this statement.

   Explain your answer, using an example from your investigation about mutations in genes.

4. Creativity is an important characteristic for a scientist to have.

   a. I agree with this statement.
   b. I disagree with this statement.

   Explain your answer, using an example from your investigation about mutations in genes.

5. Scientists often observe events that happen over different time scales; some events are very quick and others take many years. Explain why understanding different time scales is important, using an example from your investigation about mutations in genes.

6. The relationship between the structure and function of organisms' features is an important area of study in science. Discuss why it is important for scientists to understand this relationship, using an example from your investigation about mutations in genes.

Application Lab

LAB 16

Lab 16. Mechanisms of Inheritance: How Do Fruit Flies Inherit the Sepia Eye Color Trait?

Introduction

In the 1800s, farmers often cross-pollinated specific types of plants or mated livestock with specific traits in an effort to produce offspring with more desirable traits. Their selective breeding process, however, did not always lead to the desired outcome. These farmers were often unsuccessful in their attempts to produce offspring with specific traits because they did not understand the mechanisms that control how traits are passed down from parent to offspring. The inheritance of traits also baffled the scientists of that time, until Gregor Mendel was able to explain the rules that govern heredity in 1865.

Mendel was able to explain how and why specific traits are passed down from generation to generation by breeding pea plants. He first cross-pollinated individual pea plants with specific traits and documented the traits of their offspring. He then cross-pollinated individual offspring and documented the traits that shown up in the next generation. By doing this, he was able to identify patterns in the ways traits are passed down from one generation to the next. He then developed several rules that he could use to explain the patterns he uncovered in the ways traits are inherited; these rules are now called *Mendel's laws* and can be summarized as follows:

- Inheritable units called *genes* determine traits.
- A gene comes in different versions called *alleles*.
- An organism carries two alleles for each trait.
- The two alleles segregate during *gamete* (a reproductive cell such as a sperm or an egg) production, so each gamete only carries one allele for a specific trait.
- When gametes unite during fertilization, each contributes its allele, so offspring inherit one allele from each parent.

There are several different models of inheritance that are based on Mendel's laws. These models include dominant-recessive, incomplete dominance, and codominance. What makes these models of inheritance different from each other is how each one describes the interaction or behavior of the alleles once they have been passed down from parent to offspring. The *dominant-recessive model of inheritance* suggests that when an individual inherits two alleles and the two alleles differ, then one is fully expressed and determines the trait (this version of the gene is called the dominant allele) while the other one has no noticeable effect (this version of the gene is called the recessive allele). The *incomplete dominance model of inheritance* suggests that the interaction that occurs between two different alleles results in a hybrid with an appearance somewhere between the phenotypes of the two parental varieties. The *codominance model of inheritance* is similar to the incomplete dominance model, but in the codominance model both alleles affect the phenotype of the individual in separate and distinguishable ways.

It is often difficult, however, to determine which of these three models of inheritance best explains how a specific trait is inherited. To illustrate this point, you will be studying the inheritance of a trait in fruit flies (*Drosophila melanogaster*). Fruit flies are very common. Most fruit flies have six legs, two wings, and two antennae (see Figure L16.1). Most fruit flies also have an orange-yellow body and red eyes. Scientist call flies with these traits the "wild type." Every once in a while, however, you might see a fruit fly with sepia (brown) or white eyes. In this investigation, you will need to determine how the sepia eye color is inherited.

FIGURE L16.1
Male and female fruit flies

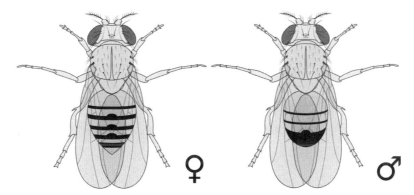

Your Task

Use a computer simulation that allows you to "order" fruit flies with specific traits from a supply company and then "breed" them to see how the sepia eye color trait is passed down from parent to offspring. From there, you will need to decide how the allele for sepia eye color interacts with other alleles for eye color (such as the allele for red eyes). You will then decide which model of inheritance (dominant-recessive, incomplete dominance, or codominance) best explains how the sepia eye color trait is inherited in fruit flies.

The guiding question of this investigation is, **How do fruit flies inherit the sepia eye color trait?**

LAB 16

Materials

You will use an online simulation called *Drosophila* to conduct your investigation. You can find the simulation by going to the following website: *www.sciencecourseware.org/vcise/drosophila*.

Safety Precautions

Follow all normal lab safety rules.

Investigation Proposal Required? ☐ Yes ☐ No

Getting Started

Your teacher will show you how to use the *Drosophila* online simulation before you begin designing your investigation. The first step in the investigation is to learn how the sepia eye color trait is passed from one generation to the next. To accomplish this task, you must determine what type of data you need to collect using the online simulation, how you will collect these data, and how you will analyze the data. To determine *what type of data you will need to collect*, think about the following questions:

- What types of flies will you need to work with during your investigation (e.g., males or females, flies with red eyes or flies with sepia eyes)?
- What type of measurements or observations will you need to record during your investigation?
- How will you identify a pattern in the way the sepia eye color trait is inherited?

To determine *how you will collect your data*, think about the following questions:

- How many times will you need to breed the flies?
- How many generations of flies will you need to follow?
- How often will you collect data and when will you do it?
- How will you keep track of the data you collect and how will you organize the data?

To determine *how you will analyze your data*, think about the following questions:

- How will you determine if the results of your cross tests match your predictions?
- What type of graph could you create to help make sense of your data?

The last step in this investigation is to test the model of inheritance that you think best explains the inheritance of the sepia eye color trait. To accomplish this goal, you can use the simulation to determine if you can use your model to make accurate predictions about how the sepia eye color trait will be passed down from one generation to the next. If you can use your model to make accurate predictions about how the traits of the flies are inherited, then you will be able to generate the evidence you need to convince others that the conceptual model you decided to use is the most appropriate one.

Connections to Crosscutting Concepts, the Nature of Science, and the Nature of Scientific Inquiry

As you work through your investigation, be sure to think about

- the importance of uncovering causes for patterns observed in nature,
- how scientists develop and use explanatory models to make sense of their observations,
- the nature of theories and laws in science, and
- the difference between data and evidence in science.

Initial Argument

Once your group has finished collecting and analyzing your data, you will need to develop an initial argument. Your argument must include a claim, evidence to support your claim, and a justification of the evidence. The claim is your group's answer to the guiding question. The evidence is an analysis and interpretation of your data. Finally, the justification of the evidence is why your group thinks the evidence matters. The justification of the evidence is important because scientists can use different kinds of evidence to support their claims. Your group will create your initial argument on a whiteboard. Your whiteboard should include all the information shown in Figure L16.2.

FIGURE L16.2

Argument presentation on a whiteboard

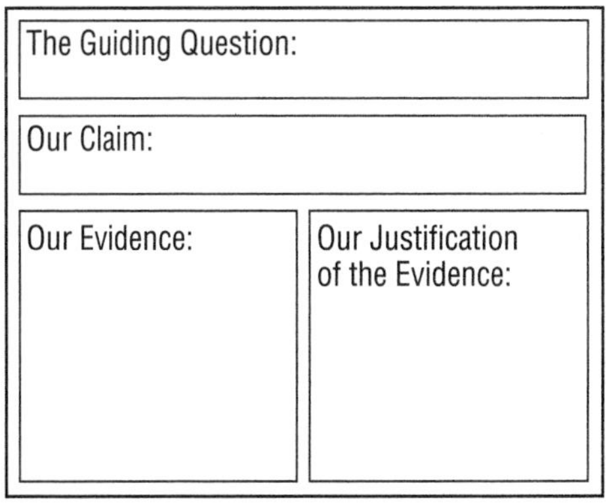

The Guiding Question:	
Our Claim:	
Our Evidence:	Our Justification of the Evidence:

Argumentation Session

The argumentation session allows all of the groups to share their arguments. One member of each group will stay at the lab station to share that group's argument, while the other members of the group go to the other lab stations one at a time to listen to and critique the arguments developed by their classmates. This is similar to how scientists present their arguments to other scientists at conferences. If you are responsible for critiquing your classmates' arguments, your goal is to look for mistakes so these mistakes can be fixed and they can make their argument better. The argumentation session is also a good time to think about ways you can make your initial argument better. Scientists must share and critique arguments like this to develop new ideas.

LAB 16

To critique an argument, you might need more information than what is included on the whiteboard. You will therefore need to ask the presenter lots of questions. Here are some good questions to ask:

- What did your group do to collect the data? Why do you think that way is the best way to do it?
- What did your group do to analyze the data? Why did your group decide to analyze it that way?
- What other ways of analyzing and interpreting the data did your group talk about?
- What did your group do to make sure that these calculations are correct?
- Why did your group decide to present your evidence in that way?
- What other claims did your group discuss before you decided on that one? Why did your group abandon those other ideas?
- How sure are you that your group's claim is accurate? What could you do to be more certain?

Once the argumentation session is complete, you will have a chance to meet with your group and revise your initial argument. Your group might need to gather more data as part of this process. Remember, your goal at this stage of the investigation is to develop the best argument possible.

Report

Once you have completed your research, you will need to prepare an investigation report that consists of three sections that provide answers to the following questions:

1. What question were you trying to answer and why?

2. What did you do during your investigation and why did you conduct your investigation in this way?

3. What is your argument?

Your report should answer these questions in two pages or less. This report must be typed and any diagrams, figures, or tables should be embedded into the document. Be sure to write in a persuasive style; you are trying to convince others that your claim is acceptable or valid!

Checkout Questions

Lab 16. Mechanisms of Inheritance: How Do Fruit Flies Inherit the Sepia Eye Color Trait?

1. In your own words, explain the difference between the dominant-recessive model of inheritance and the incomplete dominance model of inheritance.

2. Explain how a certain type of flower has two types of alleles for color, such as red and white, but produces red, white, and pink flowers.

3. Laws are theories that have been proven true.

   a. I agree with this statement.

   b. I disagree with this statement.

   Explain your answer, using an example from your investigation about mechanisms of inheritance.

4. Data are observations or measurements collected during an investigation, and evidence is analyzed data and an interpretation of the analysis.

    a. I agree with this statement.

    b. I disagree with this statement.

Explain your answer, using an example from your investigation about mechanisms of inheritance.

5. Scientists often look for patterns when they study nature. Explain why the identification of patterns in nature is so important, using an example from your investigation about mechanisms of inheritance.

6. Scientists often need to use or create conceptual models to understand a natural phenomenon. Discuss why models are important in science, using an example from your investigation about mechanisms of inheritance.

SECTION 5
Life Sciences
Core Idea 4

Biological Evolution: Unity and Diversity

Introduction Lab

LAB 17

Lab Handout

Lab 17. Mechanisms of Evolution: Why Does a Specific Version of a Trait Become More Common in a Population Over Time?

Introduction

An *ecosystem* includes all the organisms and the nonliving parts of the environment that are found in a particular area. Organisms include things such as plants, animals, fungi, and bacteria. The nonliving parts of the environment include things such as air, light, water, and minerals. The organisms found within an ecosystem depend on the nonliving components for survival. The organisms also interact with each other. For example, plants need air, light, and water to produce the food they need to survive. Animals called herbivores eat these plants. Other animals called predators eat the herbivores. Herbivores and predators also need water to drink and air to breathe in order to survive. All the living and nonliving parts of the environment therefore function as a system. A change in one part of the system will, as a result, affect the other parts of the system. For example, a drought could reduce the number of plants in a particular area. A decrease in the number of plants will result in less food for the herbivores. When these animals do not have enough food to eat, some will starve. The predators will then not have enough food to eat.

Organisms often have adaptations that allow them to function in a specific ecosystem. An adaptation can be a physical feature that helps an organism to survive. Katydids, for example, are insects that look like leaves (Figure L17.1), and their unique appearance helps them to avoid predators. An adaptation can also be something that an organism is able to do that helps it survive in a specific environment. The creosote bush (Figure L17.2), for example, reduces competition for nutrients and water by producing a toxin that prevents other plants from growing near it. Biologists define an *adaptation* as a version of a trait that is common in a population because it provides some improved function over other versions of that trait.

Organisms that live in different ecosystems tend to have different adaptations. For example, a population of herbivores that lives in an ecosystem with a lot of predators will have different adaptations than

FIGURE L17.1
A katydid

Katydids are insects that look like a leaf. The appearance of the insect is an adaptation.

FIGURE L17.2
A creosote bush

The creosote bush is a desert-dwelling plant that produces a toxin that prevents the growth of other plants.

a population of herbivores that lives in an ecosystem with few or no predators. Similarly, a herbivore population that lives in an ecosystem that gets very little rain will have different adaptations than a herbivore population that lives in an ecosystem that gets a lot of rain. It is therefore important for biologists to understand why specific versions of a trait become more or less common in a population and how changes in an ecosystem will affect the characteristics of the organisms in it. In this investigation, you will examine how a specific trait in a simulated population of bugs changes over time in two different environments. You will then develop a conceptual model that can be used to explain why a certain version of a trait becomes more common in a population over several generations.

Your Task

Use a computer simulation to explore how the frequency of different versions of the body color trait in a population of bugs changes over time in two different environments. You will then develop a model that can be used to explain why a certain version of a trait becomes more common in a population over several generations.

The guiding question of this investigation is, **Why does a specific version of a trait become more common in a population over time?**

Materials

You will use an online simulation called *Bug Hunt Camouflage* to conduct your investigation. You can find the simulation by going to the following website: *http://ccl.northwestern. edu/netlogo/models/BugHuntCamouflage*.

Safety Precautions

Follow all normal lab safety rules.

Investigation Proposal Required?   ☐ Yes   ☐ No

Getting Started

The first step in developing your model will be to use the *Bug Hunt Camouflage* simulation to explore how the frequency of different versions of the body color trait in a simulated population of bugs changes over time in two different environments. The bugs that make up the simulated population belong to the same species but are different colors. There are, as a result, individual bugs with different versions of the body color trait in the simulated environment. The color of each bug in this simulation is described in terms of its hue, saturation, and brightness (HSB). Hue is a color such as red, green, or blue and is given a value ranging from 0 to 255 in this simulation. Saturation is the purity or richness of a color and ranges from 0 (gray) to 255 (colorful) in the simulation. Brightness is the intensity of the color and, like saturation, ranges from 0 (dark) to 255 (bright) in the simulation.

Remember, all of the bugs in the simulated environment are from the same species, even though they look different.

In this simulation, you will act as the predator. You can eat the bugs (your prey) by clicking on them. When a bug is eaten, it is replaced through reproduction by another bug in the simulated ecosystem. The new bug will often (but not always) have the same color as the parent bug. The simulation provides information about the total number of bugs that you have caught since the simulation started, the current color composition of the bug population (in terms of HSB), and how the average color of the bugs in the population has changed over time. You can also change the number of bugs that are in the environment at any given time (carrying capacity) and the environment type (glacier, beach, poppy field). There are several other factors, such as bug size, that you can adjust as part of the simulation. Figure L17.3 illustrates the factors in the simulation.

FIGURE L17.3 _____

A screen shot from the *Bug Hunt Camouflage* simulation

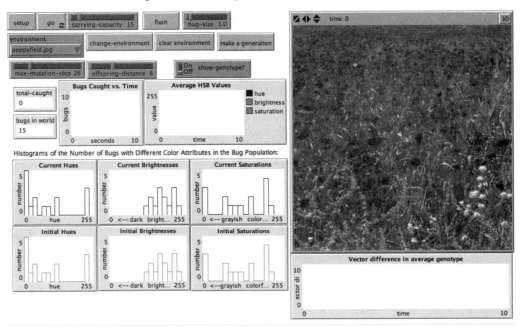

To explore how the frequency of different versions of the body color trait in the population of bugs changes over time in two different environments, you must determine what type of data you need to collect and how you will collect it using the *Bug Hunt Camouflage* simulation. You also need to determine how you will analyze the data once it has been collected. To determine *what type of data you need to collect*, think about the following questions:

- What will you do to track how the body color trait in the bug population changes over time?
- What will serve as your dependent variable (e.g., average HSB value, current hue, current brightness, number of bugs caught)?
- What type of measurements or observations will you need to record during your investigation?

To determine *how you will collect your data*, think about the following questions:

- What will serve as a control condition?
- What types of treatment conditions will you need to set up and how will you do it?
- How many trials will you need to conduct?
- How long will you need to run the simulation during each trial (e.g., for three minutes or until 60 bugs are caught)?
- How often will you record an observation or collect a measurement?
- When will you make your observations or make a measurement?
- How will you keep track of the data you collect and how will you organize it?

To determine *how you will analyze your data*, think about the following questions:

- How will you determine if there is a difference between the different treatment conditions and the control condition?
- What type of calculations will you need to make?
- What type of graph could you create to help make sense of your data?

Once you have collected and analyzed your data, your group will need to develop a conceptual model to explain why a specific version of the body color trait becomes more common in the population over several generations. Your model, however, should also be able to explain how traits in other populations of organisms can change over time. It will therefore be important for you to think about how your model could be used to explain a wide range of situations and not just what happened in your investigation.

The last step in this investigation is to test your model. To accomplish this goal, you can use a third environment in the *Bug Hunt Camouflage* simulation to determine if you can use your model to make accurate predictions about how the bug color trait changes over several generations under different conditions. If you can use your model to make accurate predictions about how the traits of the bugs in the population change in a new environment, then you will be able to generate the evidence you need to convince others that the conceptual model you developed is valid.

LAB 17

Connections to Crosscutting Concepts, the Nature of Science, and the Nature of Scientific Inquiry

As you work through your investigation, be sure to think about

- the importance of looking for patterns in nature,
- the importance of developing explanations for a natural phenomenon,
- how scientists create and use models to understand a natural phenomenon,
- the different types of methods that scientists use to answer questions, and
- the important role that imagination and creativity play in science.

Initial Argument

Once your group has finished collecting and analyzing your data, you will need to develop an initial argument. Your argument must include a claim, evidence to support your claim, and a justification of the evidence. The claim is your group's answer to the guiding question. The evidence is an analysis and interpretation of your data. Finally, the justification of the evidence is why your group thinks the evidence matters. The justification of the evidence is important because scientists can use different kinds of evidence to support their claims. Your group will create your initial argument on a whiteboard. Your whiteboard should include all the information shown in Figure L17.4.

FIGURE L17.4 _____

Argument presentation on a whiteboard

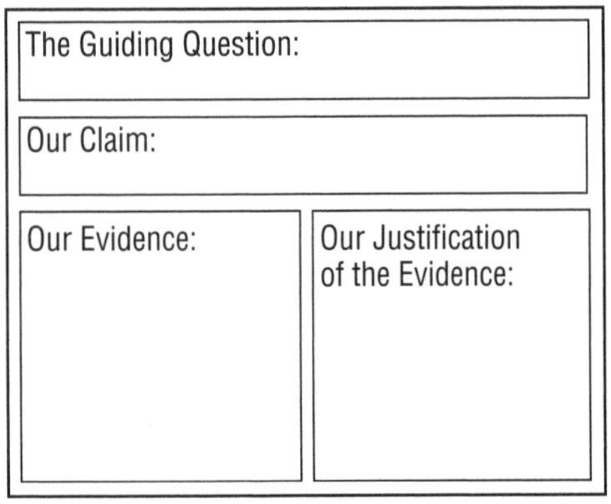

The Guiding Question:	
Our Claim:	
Our Evidence:	Our Justification of the Evidence:

Argumentation Session

The argumentation session allows all of the groups to share their arguments. One member of each group will stay at the lab station to share that group's argument, while the other members of the group go to the other lab stations one at a time to listen to and critique the arguments developed by their classmates. This is similar to how scientists present their arguments to other scientists at conferences. If you are responsible for critiquing your classmates' arguments, your goal is to look for mistakes so these mistakes can be fixed and they can make their argument better. The argumentation session is also a good time to think about ways you can make your initial argument better. Scientists must share and critique arguments like this to develop new ideas.

To critique an argument, you might need more information than what is included on the whiteboard. You will therefore need to ask the presenter lots of questions. Here are some good questions to ask:

- What did your group do to collect the data? Why do you think that way is the best way to do it?

- What did your group do to analyze the data? Why did your group decide to analyze it that way?

- What other ways of analyzing and interpreting the data did your group talk about?

- What did your group do to make sure that these calculations are correct?

- Why did your group decide to present your evidence in that way?

- What other claims did your group discuss before you decided on that one? Why did your group abandon those other ideas?

- How sure are you that your group's claim is accurate? What could you do to be more certain?

Once the argumentation session is complete, you will have a chance to meet with your group and revise your initial argument. Your group might need to gather more data as part of this process. Remember, your goal at this stage of the investigation is to develop the best argument possible.

Report

Once you have completed your research, you will need to prepare an investigation report that consists of three sections that provide answers to the following questions:

1. What question were you trying to answer and why?

2. What did you do during your investigation and why did you conduct your investigation in this way?

3. What is your argument?

Your report should answer these questions in two pages or less. The report must be typed, and any diagrams, figures, or tables should be embedded into the document. Be sure to write in a persuasive style; you are trying to convince others that your claim is acceptable or valid!

LAB 17

Lab 17. Mechanisms of Evolution: Why Does a Specific Version of a Trait Become More Common in a Population Over Time?

Snowshoe hares live in the boreal forests of Alaska, Washington, Idaho, Montana, and Canada. In winter, they grow long white guard hairs that match the snow (see the figure on the left, below). In summer, they shed their white guard hairs and have mostly rusty brown coats that blend in with trees and soil (see the figure on the right, below). Snowshoe hares are able to hide from predators (including lynx, coyotes, foxes, wolves, and birds of prey) because they are able to blend into their surroundings.

A snowshoe hare with white fur

A snowshoe hare with brown fur

The signal for a hare to shift coat color comes from the pineal gland in the brain, which senses changes in daylight length. When the days of fall get shorter, it triggers the coat color to change from brown to white. When the days get longer in the spring, the white hairs begin to shed. Usually, shorter days correspond with colder temperatures and more snowfall, so the snowshoe hare is usually white when the ground is covered with snow.

Unfortunately, the average temperature in Alaska, Washington, Idaho, Montana, and Canada has increased over the last decade and the ground is not covered in snow until well into the winter. The snowshoe hare, however, still changes color regardless of when there is snow on the ground because the shift in coat color is triggered by daylight length rather than temperature. As a result, many snowshoe hares turn white before it snows and these white hares tend to stand out against the brown background of trees and soil. Biologists have observed that the population of snowshoe hares found in these geographic areas is getting smaller because predators are catching more and more hares each fall. However, biologists also predict that the snowshoe hare population will adapt to this change in the environment.

1. Use what you have learned about how populations evolve over time to explain how this snowshoe hare population could adapt to warmer temperatures.

2. All scientists use the same method to test their ideas.

   a. I agree with this statement.

   b. I disagree with this statement.

   Explain your answer, using an example from your investigation about the mechanisms of evolution.

3. Scientists do not need to be creative or have a good imagination to be successful in science.

   a. I agree with this statement.

   b. I disagree with this statement.

   Explain your answer, using an example from your investigation about the mechanisms of evolution.

4. Scientists often attempt to identify patterns in nature. Explain why the identification of patterns is useful in science, using an example from your investigation about the mechanisms of evolution.

5. An important goal in science is to develop explanations for natural phenomena. Explain why the development of explanations is so important in science, using an example from your investigation about the mechanisms of evolution.

6. Scientists often attempt to develop models of systems in order to study them. Explain why developing a model of a system is useful in science, using an example from your investigation about the mechanisms of evolution.

Application Labs

Lab Handout

Lab 18. Environmental Change and Evolution: Which Mechanism of Microevolution Caused the Beak of the Medium Ground Finch Population on Daphne Major to Increase in Size From 1976 to 1978?

Introduction

Bacteria have developed resistance to antibiotics over time. A pesticide that was once highly effective at killing mosquitoes no longer works. House sparrows that live in the northern United States and Canada are larger-bodied than the ones that live in the southern United States and Mexico. These cases are all examples of *microevolution*, or evolutionary change on a small scale. Microevolution occurs within a population. A population is a group of organisms that live in the same area and mate with each other. Biologists define microevolution as a change in the frequency of one or more genes within a population over time. As specific genes within a population become more or less common over time, the traits that are associated with those genes will also change. There are four basic mechanisms that drive microevolution.

The first mechanism of microevolution is a genetic *mutation*. A mutation in a gene can result in an individual having a new version of a trait. The individual with the new gene can then have offspring with the same gene. The new gene could then become more common in a population over time. However, since mutations are rare and only happen in individuals, this process alone cannot result in a big change in the frequency of a gene within a population in only one or two generations.

The second mechanism of microevolution is the process of *migration*. Individuals can either join a population (immigration) or leave a population (emigration). A specific version of a gene will become less common within a population when several individuals with that gene leave the population, and a specific version of a gene will become more common within a population when several individuals with that gene join the population. The migration of a large number of individuals into or out of a population can therefore result in a dramatic shift in the frequency of a gene within a population in a relatively short period of time.

The third mechanism of microevolution is *natural selection*, which occurs when (a) there is variation in a trait among the individuals that make up a population, (b) the trait is determined by one or more genes, (c) the trait affects survival and/or ability to reproduce, and (d) individuals who reproduce pass on their genes to the next generation. The frequency of a gene in any given generation, as a result, reflects the traits and genes of the individuals that were able to survive long enough to reproduce in the previous generation. Over time, genes that determine traits that are associated with an increased chance of survival and successful reproduction will become more common in a population, and genes that determine traits that decrease an individual's chance of survival or reproduction will become less common.

The fourth, and final, mechanism of microevolution is *genetic drift*. In any generation, some individuals may, just by chance, survive longer or leave behind more offspring than other individuals. The frequency of a gene in the next generation will therefore reflect the genes and traits of these lucky individuals rather than individuals with traits that are advantageous in terms of survival or reproduction.

It is often difficult to determine which of these four mechanisms is responsible for an evolutionary change within a population. To illustrate this point, you will be studying a population of birds called the medium ground finch (*Geospiza fortis*) that lives in the Galápagos Islands, an archipelago made up of a small group of islands located 600 miles off the coast of mainland Ecuador in South America (see Figure L18.1). There is a small island in the Galápagos called Daphne Major (see Figure L18.2).

FIGURE L18.1
The Galápagos archipelago

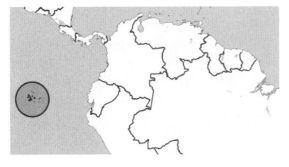

FIGURE L18.2
Daphne Major

Biologists Peter and Rosemary Grant have been studying the medium ground finch population on Daphne Major since 1974. They travel to Daphne Major every summer to study these birds. They capture, tag, and measure the physical characteristics of every bird on the island. They also keep track of the ones that die. Finally, and most importantly, they keep track of when a bird breeds, how many offspring it produces, and how many of those offspring survive long enough to breed.

In the summer of 1976, there were 751 finches on Daphne Major when the Grants left the island. The 1976 medium ground finch population had an average beak depth of 9.65 mm and an average beak length of 10.71 mm. In 1977 a severe drought began, and only 20 mm of rain fell on the island over the entire year. Much of the plant life on the island withered and died. The medium ground finches on Daphne Major, as a result, struggled to find food, and the population quickly decreased in size. By the end of 1978, there were only 90 finches left on the island. When the Grants returned to Daphne Major in 1978 to study the character- istics of the finch population, they made an unexpected discovery. They found that the average size of the beak for the medium ground finch on this island had increased. The 1978 population of the medium ground finch population on Daphne Major had an average beak depth of 10.55 mm and an average beak length of 11.61 mm, which was almost a full mm thicker and longer than the 1976 population. The beak of the medium ground finch population had clearly evolved in only two years.

The dramatic increase in the size of the medium ground finch beak was a clear example of microevolution. The Grants therefore wanted to determine which mechanism of microevolution caused the dramatic change in beak size. After they had analyzed the data that they had collected from 1976 to 1978, the Grants proposed that natural selection was the mechanism that caused the beak of the medium ground finch to increase in size. Some scientists, however, thought that this explanation was unacceptable because the change in the trait happened in only two years, and they viewed natural selection as a slow and gradual process. These scientists suggested that a better explanation for the increase in beak size was migration or genetic drift. In this investigation, you will use the data that the Grants collected on Daphne Major to determine which of these three explanations is the most valid or acceptable.

Your Task

Use the Grant's finch data set and what you know about migration, natural selection, and genetic drift to determine which of these three mechanisms of microevolution caused the average size of the medium ground finch beak to increase from 1976 to 1978.

The guiding question of this investigation is, **Which mechanism of microevolution caused the beak of the medium ground finch population on Daphne Major to increase in size from 1976 to 1978?**

Materials

You will use an Excel file called Finch Data during this investigation. The *Finch Data.xls* file is available at *www.nsta.orgadi-lifescience.aspx*.

Safety Precautions

Follow all normal lab safety rules.

Investigation Proposal Required?   ☐ Yes   ☐ No

Getting Started

You will need to examine the characteristics of the medium ground finch on Daphne Major before and after the drought of 1977 in order to answer the guiding question for this investigation. Luckily, we know a lot about the physical characteristics of all the medium ground finches found on Daphne Major.

Environmental Change and Evolution

Which Mechanism of Microevolution Caused the Beak of the Medium Ground Finch Population on Daphne Major to Increase in Size From 1976 to 1978?

The medium ground finch is a small brown bird (see Figure L18.3). Their brown color helps them blend into their surroundings and avoid the owls that live on the island. (Owls eat small birds.) As with any species, no two medium ground finches are exactly alike. Medium ground finches weigh between 12 and 17 grams and have wings that range in size from 60 mm to 70 mm. These birds also have small beaks. The beak of a medium ground finch ranges in size from 8 mm to 13 mm. The medium ground finch eats seeds (which they must crack open before eating) and the occasional insect.

You may also need to examine the characteristics of the plant life found on Daphne Major before, during, and after the drought of 1977. There are two species of seed-producing plants on Daphne Major: *Tribulus terrestris* (puncturevine) and *Portulaca oleracea* (purslane). The *Tribulus* plants produce large, hard seeds (Figure L18.4) and the *Portulaca* plants produce small, soft seeds (Figure L18.5). Medium ground finches tend to eat seeds from the *Portulaca* plants because they are soft and easy to get.

FIGURE L18.3

A medium ground finch

FIGURE L18.4

Seeds produced by the *Tribulus* plant

FIGURE L18.5

Seeds produced by the *Portulaca* plant

You will be given the observations and measurements collected by the Grants. These data have been entered into an Excel spreadsheet. The spreadsheet will make it easier for you to analyze all the available data. To answer the guiding question for this investigation, however, you must determine what type of data you will need to examine and how you will analyze it. To determine *what data you will need to examine and how you will analyze these data*, think about the following questions:

- What would you expect to see if the change in beak size in the 1976 and 1978 populations of the medium ground finch was caused by migration? Natural selection? Genetic drift?
- What types of comparisons will you need to make between the two populations to test each of the three explanations?
- Are there trends or relationships that you will need to look for in the data?
- Are there other factors that may help you test each explanation?

Connections to Crosscutting Concepts, the Nature of Science, and the Nature of Scientific Inquiry

As you work through your investigation, be sure to think about

- the important role that conceptual models play in science,
- the relationship between structure and function in nature,
- the different types of methods that scientists use to answer questions, and
- the difference between laws and theories in science.

Initial Argument

Once your group has finished collecting and analyzing your data, you will need to develop an initial argument. Your argument must include a claim, evidence to support your claim, and a justification of the evidence. The claim is your group's answer to the guiding question. The evidence is an analysis and interpretation of your data. Finally, the justification of the evidence is why your group thinks the evidence matters. The justification of the evidence is important because scientists can use different kinds of evidence to support their claims. Your group will create your initial argument on a whiteboard. Your whiteboard should include all the information shown in Figure L18.6.

FIGURE L18.6 _____

Argument presentation on a whiteboard

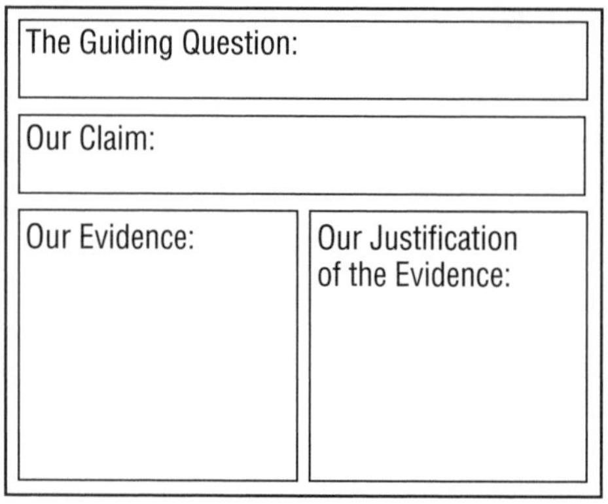

The Guiding Question:	
Our Claim:	
Our Evidence:	Our Justification of the Evidence:

Argumentation Session

The argumentation session allows all of the groups to share their arguments. One member of each group will stay at the lab station to share that group's argument, while the other members of the group go to the other lab stations one at a time to listen to and critique the arguments developed by their classmates. This is similar to how scientists present their arguments to other scientists at conferences. If you are responsible for critiquing your classmates' arguments, your goal is to look for mistakes so these mistakes can be fixed and they can make their argument better. The argumentation session is also a good time to think about ways you can

Environmental Change and Evolution

Which Mechanism of Microevolution Caused the Beak of the Medium Ground Finch Population on Daphne Major to Increase in Size From 1976 to 1978?

make your initial argument better. Scientists must share and critique arguments like this to develop new ideas.

To critique an argument, you might need more information than what is included on the whiteboard. You will therefore need to ask the presenter lots of questions. Here are some good questions to ask:

- What did your group do to analyze the data? Why did your group decide to analyze it that way?
- What other ways of analyzing and interpreting the data did your group talk about?
- Why did your group decide to present your evidence in that way?
- Why did your group abandon the other explanations?
- How sure are you that your group's claim is accurate? What could you do to be more certain?

Once the argumentation session is complete, you will have a chance to meet with your group and revise your initial argument. Your group might need to gather more data as part of this process. Remember, your goal at this stage of the investigation is to develop the best argument possible.

Report

Once you have completed your research, you will need to prepare an investigation report that consists of three sections that provide answers to the following questions:

1. What question were you trying to answer and why?
2. What did you do during your investigation and why did you conduct your investigation in this way?
3. What is your argument?

Your report should answer these questions in two pages or less. The report must be typed, and any diagrams, figures, or tables should be embedded into the document. Be sure to write in a persuasive style; you are trying to convince others that your claim is acceptable or valid!

Checkout Questions

Lab 18. Environmental Change and Evolution: Which Mechanism of Microevolution Caused the Beak of the Medium Ground Finch Population on Daphne Major to Increase in Size From 1976 to 1978?

Use the following information to answer questions 1–3.

The beach mouse (*Peromyscus polionotus*), shown in the figure below, is a small rodent found in the southeastern United States. It lives primarily in old fields and on white sand beaches. The fur of the beach mouse ranges from dark brown to very light brown. The darkest-color mice tend to live inland, and the lighter-color mice tend to live on light sand beaches.

A dark brown beach mouse

Some scientists think the trend in the coloration of the beach mouse is due to natural selection, and others think it is due to genetic drift.

1. Describe the process of natural selection, and explain how this process could result in darker-color mice living inland and lighter-color mice living on light sand beaches.

2. Describe the process of genetic drift and explain how this process could result in darker-color mice living inland and lighter-color mice living on light sand beaches.

3. Describe a test that you could conduct to determine if pattern in mouse coloration is due to natural selection or genetic drift.

4. Scientists often use existing models or develop a new model to help understand a system. Explain why models are useful in science, using an example from your investigation about environmental change and evolution.

5. The structures that make up an organism's body are not related to the functions they perform.

   a. I agree with this statement.
   b. I disagree with this statement.

   Explain your answer, using an example from your investigation about environmental change and evolution.

6. A scientific law describes the behavior of a natural phenomenon, and a scientific theory is a well-substantiated explanation of some aspect of the natural world.

   a. I agree with this statement.
   b. I disagree with this statement.

   Explain your answer, using an example from your investigation about environmental change and evolution.

7. There is no universal step-by-step scientific method that all scientists follow; rather, the choice of method depends on the objectives of the research. Explain why scientists need to use different types of methods to answer different types of questions, using an example from your investigation about environmental change and evolution.

Lab Handout

Lab 19. Phylogenetic Trees and the Classification of Fossils: How Should Biologists Classify the Seymouria?

Introduction

Biologists use *phylogenetic trees* to represent evolutionary relationships between species. A phylogenetic tree is a branching diagram that shows how various species are related to each other based on similarities and differences in their physical and/or genetic characteristics. The root of a phylogenetic tree represents a common ancestor, and the tips of the branches represent the descendants (Figure L19.1).

FIGURE L19.1 _____

Components of a phylogenetic tree

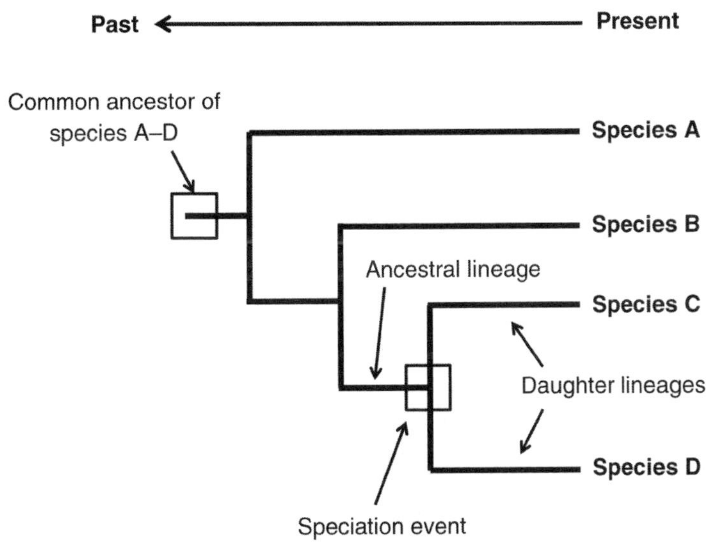

As you move from the tips to the root of the tree, you are moving backward in time. The forks in the tree represent *speciation events*. When a speciation event occurs, a single species (or an *ancestral lineage*) gives rise to two new species (or two *daughter lineages*). Each species in a phylogenetic tree has a part of its history that is unique to that species and parts that are shared with other species (Figure L19.2, p. 170). Similarly, each species has ancestors that are unique to that species and ancestors that are shared with other species (see species C and D compared with species A and B in Figure L19.2).

FIGURE L19.2

Evolutionary history in a phylogenetic tree

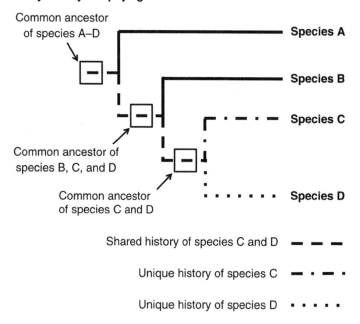

A *clade* is a grouping of species that includes a common ancestor and all the descendants (living and extinct) of that ancestor. Clades are nested within one another—biologists call this a *nested hierarchy*. A clade may include thousands of species or just a few. Some examples of clades at different levels are marked in the phylogenetic tree shown in Figure L19.3. Notice how clades are nested within larger clades.

FIGURE L19.3

Clades with a phylogenetic tree

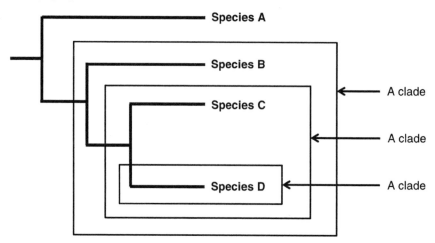

To create a phylogenetic tree, biologists collect data about the characters of organisms. Characters are heritable traits, such as physical features or genetic sequences, which can be compared across organisms. Biologists begin by examining representatives of each lineage to learn about their physical features, and they tend to look for specific features called *shared derived characters* when they create a phylogenetic tree. A shared derived character is one that developed or appeared at some point in the evolutionary history of an organism and is shared by other closely related organisms but not by distantly related organisms. Biologists then use these features to group the organisms into less and less inclusive clades.

In addition to using phylogenetic trees to represent evolutionary relationships between organisms, biologists also use them as a system of classification. The phylogenetic classification system classifies species by clades rather than assigning every one to a kingdom, phylum, class, order, family, and genus like the Linnaean system of classification. Figure L19.4 provides the evolutionary history of some major types of vertebrates and includes the names of various clades that are used to classify vertebrate species.

FIGURE L19.4

Vertebrate clades

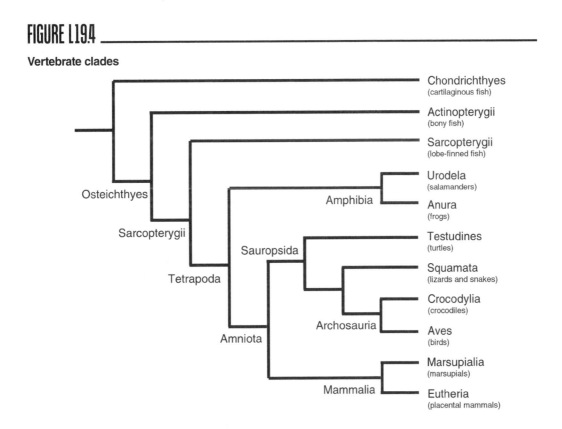

When biologists find a fossil of an extinct organism, they can use phylogenetic trees to understand how it is related to other organisms and to help classify it. The classification of an extinct organism is often difficult, however, because fossils only provide limited

LAB 19

FIGURE L19.5
Seymouria

information about the characteristics of the organism. To illustrate this point, you will attempt to classify an organism called the Seymouria (*Seymouria baylorensis*; Figure L19.5). The first fossil of the Seymouria was found in the early 1900s in Baylor County, Texas. The Seymouria was about 60 cm long and lived during the Permian Period (between 299 and 251 million years ago) throughout North America and Europe. The Seymouria has many features that make it difficult to classify. In this investigation, you will have a chance to examine the features of the Seymouria and the features of some representative species from different clades and then attempt to classify it.

Your Task

Use skeletons from vertebrates belonging to four clades (Anura, Squamata, Aves, and Eutheria), what you know about phylogenetic classification, and the characteristics of vertebrate clades to classify the Seymouria.

The guiding question of this investigation is, **How should biologists classify the Seymouria?**

Materials

You may use any of the following materials during your investigation:

- Seymouria fossil
- Frog skeleton (Anura)
- Lizard skeleton (Squamata)
- Pigeon skeleton (Aves)
- Bat skeleton (Eutheria)
- Rat skeleton (Eutheria)

Safety Precautions

Follow all normal lab safety rules. In addition, take the following safety precautions:

1. Use caution in handling skeletons. They can have sharp edges, which can cut skin.

2. Wash hands with soap and water after completing the lab activity.

Investigation Proposal Required?   ☐ Yes   ☐ No

Getting Started

To answer the guiding question, you will need to compare and contrast the features of the five modern vertebrate skeletons with the Seymouria fossil. You will be supplied with either actual specimens or images of them. You must determine what type of data you need to collect from these specimens, how you will collect the data, and how you will analyze the data. To determine *what type of data you need to collect*, think about the following questions:

- Which of the modern animals are more closely related to each other?
- What are the characteristics that biologists use to group organisms into vertebrate clades?
- Which characteristics of the specimens will you need to examine?
- How many different characteristics of the specimens will you need to examine?

To determine *how you will collect the data*, think about the following questions:

- How will you quantify differences and similarities in specimens?
- How will you make sure that your data are of high quality?
- What will you do with the data you collect?

To determine *how you will analyze your data*, think about the following questions:

- How will you compare and contrast the various specimens?
- What type of graph or table could you create to help make sense of your data?

Connections to Crosscutting Concepts, the Nature of Science, and the Nature of Scientific Inquiry

As you work through your investigation, be sure to think about

- the importance of looking for patterns in science,
- the relationship between structure and function in nature,
- the difference between observations and inferences in science, and
- how scientific knowledge changes over time.

Initial Argument

Once your group has finished collecting and analyzing your data, you will need to develop an initial argument. Your argument must include a claim, evidence to support your claim, and a justification of the evidence. The claim is your group's answer to the guiding question. The evidence is an analysis and interpretation of your data. Finally, the justification of the evidence is why your group thinks the evidence matters. The justification of the evidence is important because scientists can use different kinds of

LAB 19

FIGURE L19.6 _____

Argument presentation on a whiteboard

The Guiding Question:	
Our Claim:	
Our Evidence:	Our Justification of the Evidence:

evidence to support their claims. Your group will create your initial argument on a whiteboard. Your whiteboard should include all the information shown in Figure L19.6.

Argumentation Session

The argumentation session allows all of the groups to share their arguments. One member of each group will stay at the lab station to share that group's argument, while the other members of the group go to the other lab stations one at a time to listen to and critique the arguments developed by their classmates. This is similar to how scientists present their arguments to other scientists at conferences. If you are responsible for critiquing your classmates' arguments, your goal is to look for mistakes so these mistakes can be fixed and they can make their argument better. The argumentation session is also a good time to think about ways you can make your initial argument better. Scientists must share and critique arguments like this to develop new ideas.

To critique an argument, you might need more information than what is included on the whiteboard. You will therefore need to ask the presenter lots of questions. Here are some good questions to ask:

- What did your group do to analyze the data? Why did your group decide to analyze it that way?
- What other ways of analyzing and interpreting the data did your group talk about?
- Why did your group decide to present your evidence in that way?
- Why did your group abandon the other explanations?
- How sure are you that your group's claim is accurate? What could you do to be more certain?

Once the argumentation session is complete, you will have a chance to meet with your group and revise your initial argument. Your group might need to gather more data as part of this process. Remember, your goal at this stage of the investigation is to develop the best argument possible.

Report

Once you have completed your research, you will need to prepare an investigation report that consists of three sections that provide answers to the following questions:

1. What question were you trying to answer and why?

2. What did you do during your investigation and why did you conduct your investigation in this way?

3. What is your argument?

Your report should answer these questions in two pages or less. The report must be typed, and any diagrams, figures, or tables should be embedded into the document. Be sure to write in a persuasive style; you are trying to convince others that your claim is acceptable or valid!

Checkout Questions

Lab 19. Phylogenetic Trees and the Classification of Fossils: How Should Biologists Classify the Seymouria?

1. What is a phylogenetic tree? What is a clade?

2. This diagram shows the evolutionary relationships among several major groups of organisms. Using your knowledge of clades, identify the pairs of organisms that are *most* closely related and the pair that is *least* closely related. How many clades are there in this diagram?

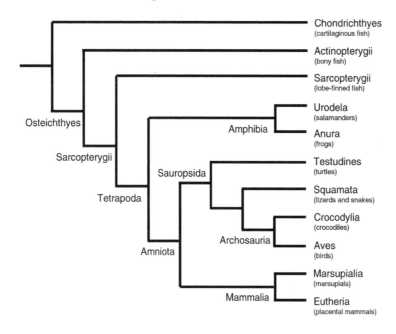

3. In science, observations and inferences are the same thing.

   a. I agree with this statement.
   b. I disagree with this statement.

   Explain your answer, using an example from your investigation about phylogenetic trees and the classification of fossils.

4. Scientific knowledge changes and develops over time.

   a. I agree with this statement.
   b. I disagree with this statement.

   Explain your answer, using an example from your investigation about phylogenetic trees and the classification of fossils.

5. Scientists often look for patterns during their investigations. Explain why patterns are important to look for, using an example from your investigation about phylogenetic trees and the classification of fossils.

6. The relationship between the structure and function of organisms' features is an important area of study in science. Discuss why it is important for scientists to understand this relationship, using an example from your investigation about phylogenetic trees and the classification of fossils.

Lab Handout

Lab 20. Descent With Modification and Embryonic Development: Does Animal Embryonic Development Support or Refute the Theory of Descent With Modification?

Introduction

One of Charles Darwin's most revolutionary ideas was that all living things are related. According to Darwin, all organisms found on Earth are related to each other because they all share a common ancestor. He argued that this common ancestor lived on Earth sometime in the distant past but is now extinct. All living things, as a result, share many of the same features, and the differences we see in organisms are simply the result of gradual modifications to these features over long periods of time.

Darwin first came to this conclusion by examining similarities and differences in the traits of closely related animals, such as the beaks of the Galápagos finches in Figure L20.1. To explain the similarities in the beaks of these birds, Darwin suggested that the birds were all the descendants of the same ancestor that originally colonized the Galápagos Islands, and the differences in their beaks were simply the result of gradual modifications in beak shape over many generations. The modifications in beak shape made the birds better adapted to survive in a particular environment. He called this theory *descent with modification*. He argued that natural selection, over time, could slowly select for or against slight variations in the basic shape of the beaks of the birds. This selection process would gradually result in some birds with thick beaks that are able to crack nuts and some birds with narrow beaks that are able to pick insects out of the bark of trees.

The theory of descent with modification can also be used to explain the existence of *homologous structures*, such as the limbs of the four animals in Figure L20.2. Homologous structures are parts of organisms that have similar components even though they may have very different functions. To explain the similar bone structure in these animals, Darwin once again argued that these animals share a common ancestor that had a limb that consisted of a humerus, an ulna, a radius, and carpals and that the observed differences in structure are simply the result of the process of natural selection. Over time, the process of natural selection gradually changed the shape of individual bones but did not completely change the basic layout of the limb. This selection process eventually resulted in whale fins and bird wings that had fingers similar to the fingers of a human or dog. These variations in structure would give their owners an advantage in a particular environment, such as the air in the case of the bird or the ocean in the case of the whale.

FIGURE L20.1

Differences in the beaks of finches found in the Galápagos Islands (Darwin 1845)

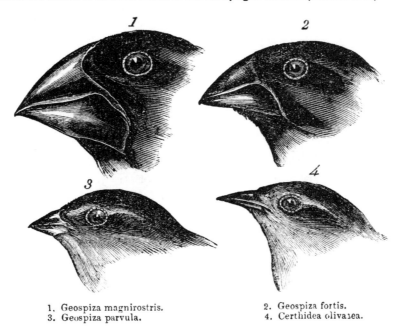

1. Geospiza magnirostris.
2. Geospiza fortis.
3. Geospiza parvula.
4. Certhidea olivaɔea.

FIGURE L20.2

Examples of homologous structures: the limbs of a human, dog, bird, and whale

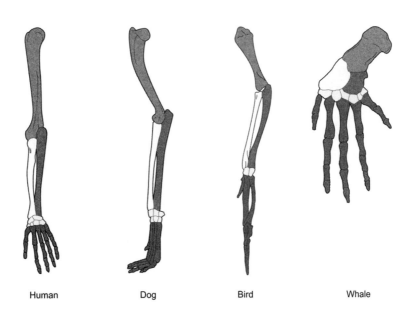

Human   Dog   Bird   Whale

LAB 20

An important principle of descent with modification is that two closely related species will have more features in common than two species that are not as closely related. For example, two species of bird will have more features in common than a bird and an amphibian. Closely related species have more features in common because they shared a common ancestor in the more recent past. More time gives natural selection more opportunities to modify features or to produce new features in each independent lineage.

The theory of descent with modification can explain many different observations, such as adaptations and homologous structures, which is one of the main criterion scientists use to evaluate the merits of a scientific theory. However, like all theories in science, the principles of descent with modification must be tested in many different ways before it can be considered valid or acceptable by the scientific community. In this investigation, you will therefore test this theory by examining the process of embryo development in animals to determine if it is consistent with the major principles of descent with modification.

Reference

Darwin, C. 1845. *Journal of researches into the natural history and geology of the countries visited during the voyage of H.M.S. Beagle round the world, under the Command of Capt. Fitz Roy, R.N.* 2nd ed. London: John Murray.

Your Task

Collect data about the embryonic development of eight different animals. Then use the data you collect to test the theory of descent with modification.

The guiding question of this investigation is, **Does animal embryonic development support or refute the theory of descent with modification?**

Materials

Your teacher will provide you with a set of the following images that can be used during your investigation (The *Lab 20 Embryo Stages.pptx* file is available at *www.nsta.org/ adi-lifescience*):

- Images of amphibian (frog) embryo development
- Images of bird (fieldfare) embryo development
- Images of bird (quail) embryo development
- Images of fish (zebrafish) embryo development
- Images of mammal (bat) embryo development
- Images of mammal (mouse) embryo development
- Images of reptile (snake) embryo development
- Images of reptile (turtle) embryo development

Safety Precautions

Follow all normal lab safety rules.

Investigation Proposal Required?   ☐ Yes   ☐ No

Getting Started

All animals begin life as a single cell. This single cell then starts to divide and becomes an embryo. As the cells continue to divide, specific structures such as organs and limbs begin to take shape. The embryo eventually starts to look more and more like the animal it will become as it changes and grows in size. Biologists have been studying the process of embryo development in animals since the 19th century. We therefore know a lot about what animal embryos look like as they develop and grow over time.

To answer the guiding question, you will need to compare and contrast the process of embryo development for eight different animals. You will be supplied with 10 embryo images for each animal. The images represent specific embryonic development milestones. You must determine what type of data you need to collect from these images, how you will collect it, and how you will analyze it. To determine *what type of data you need to collect*, think about the following questions:

- What would you expect the process of embryo development to look like in these eight different animals if they shared a common ancestor? What would it look like if they did not share a common ancestor?
- Which animals are more closely related to each other?
- Which characteristics of the embryo will you examine?
- How many different characteristics of the embryos will you need to examine?

To determine *how you will collect the data*, think about the following questions:

- How will you quantify differences and similarities in embryos?
- How will you make sure that your data are of high quality?
- What will you do with the data you collect?

To determine *how you will analyze your data*, think about the following questions:

- How will you compare and contrast the various embryos?
- What type of graph or table could you create to help make sense of your data?

LAB 20

Connections to Crosscutting Concepts, the Nature of Science, and the Nature of Scientific Inquiry

As you work through your investigation, be sure to think about

- the importance of looking for patterns in nature,
- the relationship between structure and function in nature,
- how science as a body of knowledge develops over time, and
- the different types of methods that scientists use to answer questions.

Initial Argument

Once your group has finished collecting and analyzing your data, you will need to develop an initial argument. Your argument must include a claim, evidence to support your claim, and a justification of the evidence. The claim is your group's answer to the guiding question. The evidence is an analysis and interpretation of your data. Finally, the justification of the evidence is why your group thinks the evidence matters. The justification of the evidence is important because scientists can use different kinds of evidence to support their claims. Your group will create your initial argument on a whiteboard. Your whiteboard should include all the information shown in Figure L20.3.

FIGURE L20.3 _____

Argument presentation on a whiteboard

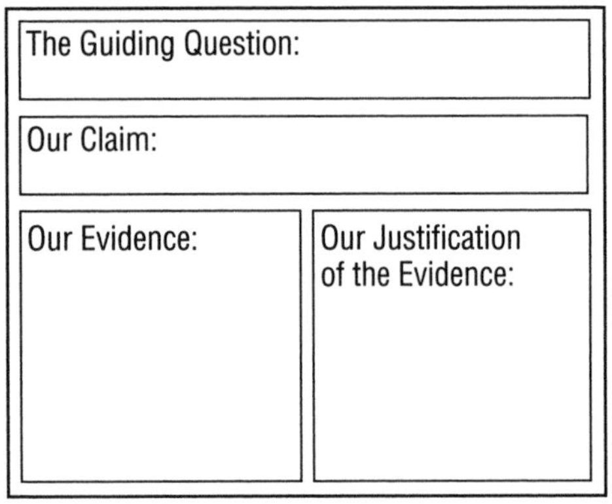

The Guiding Question:	
Our Claim:	
Our Evidence:	Our Justification of the Evidence:

Argumentation Session

The argumentation session allows all of the groups to share their arguments. One member of each group will stay at the lab station to share that group's argument, while the other members of the group go to the other lab stations one at a time to listen to and critique the arguments developed by their classmates. This is similar to how scientists present their arguments to other scientists at conferences. If you are responsible for critiquing your classmates' arguments, your goal is to look for mistakes so these mistakes can be fixed and they can make their argument better. The argumentation session is also a good time to think about ways you can make your initial argument better. Scientists must share and critique arguments like this to develop new ideas.

To critique an argument, you might need more information than what is included on the whiteboard. You will therefore need to ask the presenter lots of questions. Here are some good questions to ask:

- What did your group do to analyze the data? Why did your group decide to analyze it that way?
- What other ways of analyzing and interpreting the data did your group talk about?

- Why did your group decide to present your evidence in that way?
- Why did your group abandon the other explanations?
- How sure are you that your group's claim is accurate? What could you do to be more certain?

Once the argumentation session is complete, you will have a chance to meet with your group and revise your initial argument. Your group might need to gather more data as part of this process. Remember, your goal at this stage of the investigation is to develop the best argument possible.

Report

Once you have completed your research, you will need to prepare an investigation report that consists of three sections that provide answers to the following questions:

1. What question were you trying to answer and why?

2. What did you do during your investigation and why did you conduct your investigation in this way?

3. What is your argument?

Your report should answer these questions in two pages or less. The report must be typed, and any diagrams, figures, or tables should be embedded into the document. Be sure to write in a persuasive style; you are trying to convince others that your claim is acceptable or valid!

Checkout Questions

Lab 20. Descent With Modification and Embryonic Development: Does Animal Embryonic Development Support or Refute the Theory of Descent With Modification?

1. What are the basic principles of the theory of descent with modification?

2. Use the theory of descent with modification to explain why all mammals have the same set of bones in their limbs. The forelimbs of dogs and whales, for example, include a humerus, an ulna, a radius, and several carpals.

3. Scientific knowledge changes and develops over time.

    a. I agree with this statement.

    b. I disagree with this statement.

    Explain your answer, using an example from your investigation about descent with modification and embryonic development.

4. In science, there are usually multiple ways to investigate a question.

    a. I agree with this statement.

    b. I disagree with this statement.

    Explain your answer, using an example from your investigation about descent with modification and embryonic development.

5. Scientists often look for patterns during their investigations. Explain why patterns are important to look for, using an example from your investigation about descent with modification and embryonic development.

6. The relationship between the structure and function of organisms' features is an important area of study in science. Discuss why it is important for scientists to understand this relationship, using an example from your investigation about descent with modification and embryonic development.

IMAGE CREDITS

Lab 1

Figure L1.1: OpenStax College, Wikimedia Commons, CC BY 3.0. *http://commons.wikimedia.org/wiki/File:0315_Mitochondrion_new.jpg*

Figure L1.2: Courtesy of Jonathon Grooms.

Figure L1.3: Authors

Lab 2

Figure L2.1: User:At09kg, Wikimedia Commons, CC BY-SA 3.0. *http://commons.wikimedia.org/wiki/File:Plants.gif*

Figure L2.2: Authors

Figure L2.3: Authors

Checkout Questions figure: Authors

Lab 3

Figure L3.1: User:Pidalka44, Wikimedia Commons, Public domain. *http://commons.wikimedia.org/wiki/File:Semipermeable_membrane.png*

Figure L3.2: User:LadyofHats, Wikimedia Commons, Public domain. *http://commons.wikimedia.org/wiki/File:Osmotic_pressure_on_blood_cells_diagram-sk.svg*

Figure L3.3: Courtesy of Jonathon Grooms.

Figure L3.4: Authors

Checkout Questions figures: Authors

Lab 4

Figure L4.1: OpenStax College, Wikimedia Commons, CC BY 3.0. *http://commons.wikimedia.org/wiki/File:0312_Animal_Cell_and_Components.jpg*

Figure L4.2: Authors

Lab 5

Figure L5.1: Daniel Mayer, Wikimedia Commons, CC BY-SA 4.0, GFDL 1.2. *http://commons.wikimedia.org/wiki/File:Simple_photosynthesis_overview.svg*

Figure L5.2: Authors

Figure L5.3: Authors

Lab 6

Figure L6.1: U.S. Food and Drug Administration, Wikimedia Commons, Public domain. *http://commons.wikimedia.org/wiki/File%3AFDA_Nutrition_Facts_Label_2006.jpg*

Figure L6.2: Courtesy of Jonathon Grooms.

Figure L6.3: Authors

Lab 7

Figure L7.1: Theresa Knott, Wikimedia Commons, CC BY-SA 2.5, GFDL 1.2. *http://commons.wikimedia.org/wiki/File:Respiratory_system.svg*

Figure L7.2: User:Sansculotte, Wikimedia Commons, CC BY-SA 3.0. *http://commons.wikimedia.org/wiki/File:Grafik_blutkreislauf.jpg*

Figure L7.3: Authors

Lab 8

Checkout Questions figure: Authors

Lab 9

Figure L9.1: Bob Blaylock, Wikimedia Commons, CC BY-SA 3.0. *http://commons.wikimedia.org/wiki/File:20100911_232323_Yeast_Live.jpg*

Figure L9.2: Authors

Lab 10

Figure L10.1: Authors

Image Credits

Lab 11

Figure L11.1: Tyler Rubley, Wikimedia Commons, CC BY-SA 3.0. *http://commons.wikimedia.org/wiki/File:Mosquito_Energy_Transfer_Food_Web.pdf*

Figure L11.2: Authors

Checkout Questions figure: Authors

Lab 12

Figure L12.1: Authors

Lab 12 Reference Sheet

Nitrogen cycle: Johann Dréo, Wikimedia Commons, CC BY-SA 3.0, GFDL 1.2. *http://commons.wikimedia.org/wiki/File:Cicle_del_nitrogen_ca.svg*

Phosphorus cycle: User:Bonniemf, Wikimedia Commons, CC BY-SA 3.0. *http://commons.wikimedia.org/wiki/File:Phosphorus_cycle.png*

Lab 13

Figure L13.1: Adapted from Harry C, Wikimedia Commons, CC BY-SA 3.0. *http://commons.wikimedia.org/wiki/File:Carbon-cycle-full.jpg*

Figure L13.2: Authors

Lab 14

Figure L14.1: Blatchley, W. S. 1859-1940, Wikimedia Commons, Public domain. *http://commons.wikimedia.org/wiki/File:Dynastes_tityusBlatchleyF312A.jpg*

Figure L14.2: User:Siga, Wikimedia Commons, CC BY-SA 3.0, GFDL 1.2. *http://commons.wikimedia.org/wiki/File:Carabus_violaceus_up.jpg*

Figure L14.3: Authors

Lab 14 Reference Sheet

Harpalus affinis: a: User:©entomart, Wikimedia Commons. *http://commons.wikimedia.org/wiki/File:Harpalus_affinis01.jpg; b:* James Lindsey at Ecology of Commanster, Wikimedia Commons, CC BY-SA 3.0. *http://commons.wikimedia.org/wiki/File:Harpalus.affinis.jpg; c:* User:Futureman1199, Wikimedia Commons, CC BY-SA 3.0. *http://commons.wikimedia.org/wiki/File:Harpalus_affinis.jpg*

Cotinis mutabilis: a: User:Davefoc, Wikimedia Commons, CC BY-SA 3.0, GFDL 1.2. *http://commons.wikimedia.org/wiki/File:CotinisMutabilis_7871.JPG;* b: User:Davefoc, Wikimedia Commons, CC BY-SA 3.0, GFDL 1.2. *http://commons.wikimedia.org/wiki/File:CotinisMutabilis_7864.JPG;* c: Eugene Zelenko, Wikimedia Commons, CC BY-SA 3.0, GFDL 1.2. *http://commons.wikimedia.org/wiki/File:Cotinis_mutabilis-3.jpg.*

Leptinotarsa decemlineata: a: Fritz Geller-Grimm, Wikimedia Commons, CC BY-SA 3.0. *http://commons.wikimedia.org/wiki/File:Leptinotarsa_fg02.jpg;* b: User:Barbarossa, Wikimedia Commons, CC BY-SA 3.0, GFDL 1.2. *http://commons.wikimedia.org/wiki/File:Coloradokever_dichtbij.png;* c: Scott Bauer, USDA ARS, Wikimedia Commons, Public domain. *http://commons.wikimedia.org/wiki/File:Colorado_potato_beetle.jpg.*

Lab 15

Figure L15.1: Adapted from すじにくシチュー, Wikimedia Commons, Public domain. *http://commons.wikimedia.org/wiki/File:DNA%E3%81%AE%E4%B8%A6%E3%81%B3%E6%96%B9.png*

Figure L15.2: Jessica Reuter, Wikimedia Commons, Public domain. *http://commons.wikimedia.org/wiki/File:Central_dogma.JPG*

Figure L15.3: Courtesy of Patrick Enderle.

Figure L15.4: Authors

Lab 16

Figure L16.1: User:Madboy74, Wikimedia Commons, Public domain. *http://commons.wikimedia.org/wiki/File:Biology_Illustration_Animals_Insects_Drosophila_melanogaster.svg*

Figure L16.2: Authors

Lab 17

Figures L17.1: User:Vishalsh521, Wikimedia Commons, CC BY-SA 3.0. *http://commons.wikimedia.org/wiki/File:Katydid_india.jpg*

Figure L17.2: User:Sue in az, Wikimedia Commons, Public domain. *http://commons.wikimedia.org/wiki/File:Creosote_Larrea_tridentata.JPG*

Figure L17.3: *http://ccl.northwestern.edu/netlogo/models/BugHuntCamouflage.* BugHuntCamouflage via authors

Figure L17.4: Authors

Checkout Questions figures: A snowshoe hare with a white fur: D. Gordon E. Robertson, Wikimedia Commons, CC BY-SA 3.0. *http://commons.wikimedia.org/wiki/File:Snowshoe_Hare,_Shirleys_Bay.jpg*; A snowshoe hare with a brown fur: U.S. Fish and Wildlife Service, Wikimedia Commons, Public domain. *http://commons.wikimedia.org/wiki/File:Snowshoe_hare_eating_grass.jpg.*

Lab 18

Figures L18.1: User:TUBS, Wikimedia Commons, CC BY-SA 3.0, GFDL 1.2. *http://commons.wikimedia.org/wiki/File:Galapagos_Islands_in_South_America_(-mini_map_-rivers).svg*

Figure L18.2: Mike Weston, Wikimedia Commons, CC BY 2.0. *http://commons.wikimedia.org/wiki/File:Daphne_Major.jpg*

Figure L18.3: User:Charlesjsharp, Wikimedia Commons, CC BY-SA 3.0. *http://commons.wikimedia.org/wiki/File:Female_Galápagos_medium_ground_finch.jpg*

Figures L18.4: Forest & Kim Starr, Wikimedia Commons, CC BY 3.0. *http://commons.wikimedia.org/wiki/File:Starr_060228-6323_Tribulus_cistoides.jpg*

Figure L18.5: User:6th Happiness, Wikimedia Commons, CC BY-SA 3.0. *http://commons.wikimedia.org/wiki/File:6H-diGangi-Purslane-Seed-Pods.jpg*

Figure L18.6: Authors

Checkout Questions figure: Roger W. Barbour, United States Fish and Wildlife Service, Wikimedia Commons, Public domain. *http://commons.wikimedia.org/wiki/File:Peromyscus_polionotus_oldfield_mouse.jpg*

Lab 19

Figure L19.1: Authors

Figure L19.2: Authors

Figure L19.3: Authors

Figure L19.4: Authors

Figure L19.5: Ryan Somma, Wikimedia Commons, CC BY-SA 2.0. *http://commons.wikimedia.org/wiki/File:Seymouria.jpg*

Figure L19.6: Authors

Checkout Questions figure: Authors

Lab 20

Figure L20.1: John Gould, Wikimedia Commons, Public domain. *http://commons.wikimedia.org/wiki/File:Darwin%27s_finches_by_Gould.jpg*

Figure L20.2: Волков Владислав Петрович, Wikimedia Commons, Public domain. *http://commons.wikimedia.org/wiki/File:Homology_vertebrates.svg*

Figure L20.3: Authors